Contents

An Assessment of Renewable Energy for the UK

London: HMSO

ETSU ETSU (Energy Technology Support Unit) was established in 1974 and since that time it has become a national centre of expertise in energy technologies. It manages the UK Renewable Energy Programme for the Department of Trade and Industry and assists it in the strategic analysis of energy technology options for the UK. Through its work for the Department of Trade and Industry (DTI), the former Department of Energy, and the European Commission, ETSU has built an internationally recognised competence in both programme planning and management and in the assessment of technical, economic and environmental aspects of energy technologies. The DTI commissioned ETSU to undertake this assessment as an integral part of the Government's current review of Renewable Energy.

ETSU disseminates for the DTI a wide range of information on renewable energy sources and technologies, the bulk of which is directly related to activities carried out under the UK Renewable Energy Programme. For further information about the publications contact:

Renewable Energy Enquiries Bureau
ETSU
Harwell
Oxon
OX11 0RA

Telephone: 0235 432450

The paper used in this publication is derived from sustainable forest and produced by a totally chlorine free process.

Designed and Typeset by Design Services, Harwell

Preface

This document assesses the renewable energy technologies that may be suitable for application within the UK and provides an evaluation of their current status and potential contribution over the next three decades. It has been prepared as part of the Department of Trade and Industry's strategic review of renewable energy.

The objectives set by the Department of Trade and Industry for the Assessment were:

- to review the current status of renewable energy in the UK and abroad;
- to assess the extent to which renewable energy technologies can contribute to the energy supply of the UK;
- to consider the environmental benefits and impacts which may result from the deployment of renewable energy technologies;
- to identify the major barriers to the deployment of renewable energy and means to overcome them.

The Assessment has been conducted in conjunction with the Appraisal of UK Energy Research, Development, Demonstration and Dissemination (RDD&D) also undertaken by ETSU for the Department of Trade and Industry. The Appraisal covers all energy technologies, both supply (fossil fuelled, nuclear, renewable) and demand (domestic, industry, transport) as well as various conversion technologies of relevance to the UK. To ensure consistency, the Assessment serves as the renewable energy volume of the Appraisal. The results of the Appraisal's assessment of renewable energy RDD&D requirements and opportunities, which identify RDD&D priorities, are presented in Annex 2. The main Appraisal report, covering all energy technologies, will be published early in 1994.

Finally, as Programme Manager – Renewable Energy, I wish to add a personal comment. The work undertaken to complete this Assessment has involved many colleagues at ETSU. It has been a true team effort. I would particularly like to acknowledge here the important contributions made by those responsible for preparing the technology modules which constitute the backbone of the Assessment.

Overview	James Cavanagh
Wind	Ian Page
Hydro	John Clarke, John Robertson
Tidal	Roger Price, James Craig
Wave	John Clarke, Tom Thorpe
Photovoltaics	Adrian Cole
Photoconversion	Adrian Cole, Suzanne Evans
Active Solar	Dave Stainforth
Passive Solar Design	Adrian Cole
Geothermal Hot Dry Rocks	Paul Macdonald
Geothermal Aquifers	Paul Macdonald
Municipal and General Industrial Wastes	Anton van Santen
Landfill Gas	Keith Brown, David Maunder
Specialised Industrial Wastes	Anton van Santen
Agricultural and Forestry Wastes	Paul Maryan
Energy Crops	Paul Maryan
Advanced Conversion	Philip Guildford
RDD&D Requirements and Opportunities	Garry Staunton

The editorial team consisted of James Cavanagh, Vivien Brooks and myself. James Cavanagh acted as project manager and was responsible for the co-ordination of the work.

Eric Bevan,
Programme Manager Renewable Energy, ETSU.

1. Renewable Energy Technology Today

1.1. The Renewable Energy Technologies

Role of the Renewables Today

Interest in renewable energy has been growing steadily over the past twenty years. Today few would perceive a future without the renewables contributing to our energy provision and many believe that renewable energy will make a substantial contribution to our energy supplies in the longer term. This interest has been heightened by a number of concerns over the use of conventional energy technologies and their environmental impacts.

In the 1970s the dramatic fluctuations in the price of oil caused governments to look for ways of increasing the diversity of their energy supplies. Attention was turned towards renewable energy since replacing imported fossil fuels with indigenous renewable resources would help to achieve a more balanced energy portfolio with less risk of interruptions in supply and destabilising price rises. Many countries established programmes of research, development and demonstration for the renewables and today, after 20 years of development experience, a number of renewables which benefited from these programmes are being commercially deployed in niche markets and show prospects for wider uptake.

Description and Current Status of Technologies

In recent years increased concern about the environment, and in particular the environmental impacts of conventional energy systems – global warming, acid rain, etc. – has revitalised interest in the renewables. With little or no net emissions of polluting gases the renewables are seen as part of a solution to these problems.

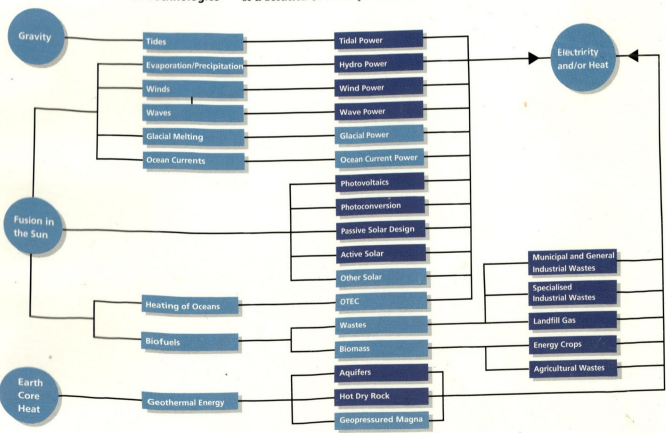

Figure 1: Renewable Energy Flowchart. Darker boxes indicate that the resource is dealt with in detail in Annex 1.

Renewable energy is the term used to cover those energy flows that occur repeatedly in the environment and can be harnessed for human benefit. The ultimate sources of most of this energy are the sun, gravity and the earth's rotation.

The technologies to harness these renewable energy flows are many and various. Brief descriptions of the renewable energy technologies that have been considered candidates for deployment in the UK are presented below. A full description of their status and prospects is given within individual modules for each technology in Annex 1.

Wind Power

Wind power is an intermittent resource which is strongly influenced by geographical effects such as the local terrain. The amount of wind energy is a strong function of the wind speed, the instantaneous power in the wind increasing as the cube of the wind speed.

Wind power has been harnessed by Man for over 2,000 years and is one of the most promising renewable energy sources for electricity generation. There are two basic design configurations – horizontal axis machines and vertical axis machines. Horizontal axis designs are at a more advanced stage of development and the evidence is increasing that they are also more cost-effective. Apart from the need to demonstrate adequate lifetimes, there is no doubt about the technical feasibility of harnessing wind power.

The existing technology offers a range of power ratings from a few kilowatts up to several megawatts. The technology is well established, with over 20,000 grid connected machines in operation world-wide. Current development work is concentrating on reliability, the further reduction of cost and noise levels, aspects of the electrical connection into the grid and overall performance. Typically about 10 to 25 machines would be deployed in a wind farm: 25 machines might produce around 20 GWh/year of electricity, enough for about 5,000 homes.

The UK has the largest wind resource in Europe. At present there is approximately 125 MW rated capacity in operation mainly in England and Wales as a result of the premium prices available through the Non-Fossil Fuel Obligation. These figures are to be compared with over 350 MW installed in Denmark, over 100 MW in the Netherlands and over 80 MW in Germany.

There are limitations on the availability of land for wind turbine sites due both to physical constraints – such as the presence of towns, villages, lakes, rivers, woods, roads and railways – and institutional constraints such as the protection of land areas designated as being of national importance. Offshore there is potentially a very large wind resource but it will require additional technology development before it can be effectively exploited and it will cost more than onshore wind.

Hydro Power

Hydro power comes from the energy available from water flowing in a river or in a pipe from a reservoir. Evidence of the use of hydro power as a source of energy has been found in primitive devices from the first century BC. During the Industrial Revolution, small-scale hydro power was commonly used to drive mills and various types of machinery. The first large-scale hydro scheme in the UK was built in Scotland in 1896.

Hydroelectric technology can be regarded as being fully commercialised. Turbine plant, engineering services and turnkey systems are sold by UK and

Typical 400 kW wind turbine at Delabole, Cornwall
(photograph courtesy of P. Edwards.)

Large-scale hydro dam, Sloy, Scotland
(courtesy Scottish Hydro-Electric plc).

overseas organisations. Numerous schemes have been built in the UK, ranging from installations producing less than 1 kW to more than 100 MW. Hydroelectric schemes fall into two broad categories – large-scale and small-scale. Large-scale schemes are considered as those with an installed capacity in excess of 5 MW and were built by the electricity utility companies. Most of the large-scale hydro capacity in the UK is installed in Scotland (1.22 GW), with a smaller amount in Wales (134 MW). The small-scale sites comprise about 58 MW in Scotland and 20 MW in England & Wales. Hydro power accounts for about 2% of the total installed generating capacity in the UK. An additional category is low-head hydro (less than 3 m which is usually small-scale). There is potential for the further development of hydroelectric power in the UK although the potential for large schemes is limited for environmental reasons.

Tidal Power

Tidal energy barrage, La Rance, France.

Tides are caused by the gravitational attraction of the moon and sun acting on the oceans of the rotating earth. The relative motions of these bodies cause the surface of the oceans to be raised and lowered periodically. In the open ocean, the maximum amplitude of the tides is about 1 m. Towards the coast, the tidal amplitudes are increased by the shelving of the sea bed and funnelling of estuaries. For example, in the Severn Estuary the maximum amplitude is about 11 m.

The energy obtainable from a tidal scheme varies with location and time. The available energy is approximately proportional to the square of the tidal range, and the output changes not only as the tide ebbs and floods each day but can vary by a factor of four over a spring-neap cycle. However, this output is exactly predictable in advance. Extraction of energy from tides is considered to be practical only at those sites where the energy is concentrated in the form of large tides and in estuaries where the geography provides suitable sites for tidal plant construction. Such sites are not commonplace, but a considerable number have been identified in the UK, which probably has the most favourable conditions in Europe for generating electricity from the tides. This is the result of an unusually high tidal range along the west coast of England and Wales, where there are many estuaries and inlets which could be exploited. On the other hand, the tidal ranges in Scotland and Northern Ireland are generally too low for economic exploitation.

A tidal barrage is a major construction project built across an estuary, consisting of a series of gated sluices and low-head turbine generators. Several locations around the world have been studied as potential barrage sites, but relatively few tidal power plants have been constructed. The first and largest (240 MW) tidal plant was built in the 1960s at La Rance in France, and has now completed more than 25 years of successful commercial operation. There is no electricity generated from tidal power in the UK.

Wave Energy

Artist's impression of a section through the Islay wave energy device.

Ocean waves are caused by the transport of energy from winds as they blow across the surface of the sea. The amount of energy transferred depends upon the speed of the wind and the distance over which it acts. As deep ocean waves suffer little energy loss, they can travel long distances if there is no intervening land mass. Therefore the western coastline of Europe has one of the largest wave energy resources in the world, being able to receive waves generated by storms throughout the Atlantic.

Wave energy is still in the RD&D phase. Currently there are two types of device known to be operating in Europe. The Norwegians have developed a tapered channel device (Tapchan) but the concept is limited to use in areas where there is a small tidal rise and fall and having suitable shoreline topography. In the UK, since 1985, the Government has funded work on a 75 kW oscillating water column device incorporating a Wells air turbine developed by the Queens University, Belfast. This device is now connected to the grid on Islay in the Inner Hebrides.

World-wide, installed devices are limited to experimental plants of less than 100 kW, including oscillating water column devices incorporated in sea defence breakwaters.

Photovoltaics

House incorporating photovoltaic cells, Milton Keynes.

Photovoltaic (PV) materials generate direct current electrical power when exposed to light. Power generation systems using these materials have the advantage of no moving parts and can be formed from thin layers (1 to 250 microns) deposited on readily available substrates such as glass. To date, the photovoltaic effect has been widely exploited where low power requirements, good solar resource and simplicity of operation outweigh the high cost of PV systems. Current applications include consumer goods, such as calculators and watches and, on a larger scale, power systems for lighting and water pumping in developing countries and in remote areas with no grid supply. Other applications include powering of "professional" systems such as remote telecommunications facilities and cathodic protection of pipelines.

There is world-wide interest in developing PV systems for future power generation because of the huge potential renewable resource available and the environmental benefits offered by a technology which avoids the emissions and pollution associated with fossil-fuelled plant. However, PV is still a relatively young technology. Much research and development will be necessary if world-wide system costs (modules and associated components) are to be reduced to acceptable levels and significant new markets are to be established.

PV could contribute to electricity supply in two ways – through the use of central PV generating plant (PV power stations) or through building-integrated systems where PV units would be located in the facades of domestic and commercial buildings. Building integrated systems could supply power for use inside the buildings for applications such as appliances, air conditioning and lighting, with any excess available for export to the grid. At present, the economic prospects for central generating plant look poor for the UK. Building-integrated systems might become an economic proposition in the future, even under UK climatic conditions, if the PV system can be incorporated into the building structure, displacing conventional building materials and components.

Photoconversion

Photoconversion is a term for various processes which convert sunlight directly into either electrical power, heat or a chemical fuel. These include photobiological, photochemical and photoelectrochemical processes. Photoconversion technology is still at the laboratory research stage. Several photoconversion processes are being investigated world-wide, but at present only two are thought to be worthy of attention as possible future energy options for the UK – electrochemical photovoltaic cells and photobiological systems to produce hydrogen.

Active Solar and Thermal Solar Power

Collectors on a swimming pool (*Courtesy Robinson's Development*).

Active solar thermal systems consist of solar collectors, which transform solar radiation into heat, connected to a heat distribution system. Due to the nature of the UK climate, such heating systems are best suited to applications at temperatures below 100°C. High temperature applications, such as thermal solar power for electricity generation, are not practical in the UK.

There is a developed technology and an existing small market for systems to supplement the heating energy demands of buildings. This market is served by a small number of manufacturers and installers, but many of the installers see solar heating as a secondary activity associated with another business, such as central heating installation.

Passive Solar Design

Passive solar design (PSD) aims to maximise free solar gains to buildings so as to reduce their energy requirements for heating or cooling and lighting. It is most effective when used with energy efficiency measures as an integral part of energy-conscious design of new buildings. However, some PSD features, such as conservatories and roof space collectors, can be retrofitted.

The concept of PSD is not new. However, its potential energy benefits, as distinct from its use for aesthetic or health reasons, have only recently become a focus of attention. To maximise these benefits in terms of the heating requirements of a building, PSD seeks to orientate and arrange glazed surfaces so as to make full use of shortwave solar radiation for heating interior spaces and to avoid heat loss resulting from siting windows on shaded walls. To cool buildings it uses solar heated air to assist natural convection, thus providing natural ventilation and cooling. For lighting, it uses glazing to reduce the need for artificial lighting whilst still maintaining a comfortable environment. More complex approaches such as mass walls, atria or conservatories are basically extensions of these simple design principles. Effective use of PSD depends on sympathetic interior design and on grouping buildings to minimise shading and gain protection from prevailing winds.

Passive solar estate in South London.

Geothermal Hot Dry Rock

There is a large amount of heat just below the earth's surface - much of it stored in low permeability rocks such as granite. This source of geothermal heat is called "hot dry rock" (HDR). Attempts to extract the heat have been based on drilling two holes from the surface. Water is pumped down one of the boreholes, circulated through the naturally occurring, but artificially dilated, fissures present in the hot rock, and returned to the surface via the second borehole. The superheated water or steam reaching the surface can be used to generate electricity or for combined heat and power systems. The two boreholes are separated by several hundred metres in order to extract the heat over a sizeable underground volume. A typical HDR power station would produce about 5 MW of electricity and be expected to operate for at least 20 years.

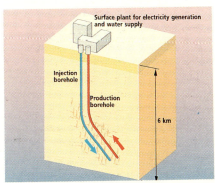

A conceptual geothermal hot dry rock scheme.

The engineering of the underground "heat exchanger" has turned out to be a formidable technical problem which has not yet been satisfactorily solved after more than ten years of intensive research in this country, the USA and elsewhere. Because of the technical difficulties, there are no commercial HDR schemes in existence anywhere in the world.

A review of the UK HDR research programme in 1990 concluded that the fundamental technical difficulties made the commercial development of HDR unrealistic in the foreseeable future and even if the technical problems were overcome, the costs of generating electricity by an HDR power station located in the UK would be many times greater than by conventional means of power generation.

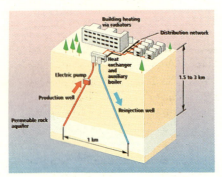
A typical geothermal aquifer scheme.

Geothermal Aquifers

Geothermal aquifers extract heat from the earth's crust through naturally occurring ground waters in porous rocks at depth. A borehole is drilled to access the hot water or steam, which is then passed through a heat exchanger located on the surface. If the temperature of the hot fluid exceeds about 150°C it can be used for generating electricity; otherwise it can be tapped as a source of warm water. In the UK, there are very few sources with temperatures above 60°C and the resource would be exploitable mainly in district heating systems or industrial processes.

The use of aquifers is well established in certain geologically favoured parts of the world, such as Iceland, Hungary, Italy, the USA and the Paris Basin of France. Some 6 GW of electrical generating capacity is currently installed overseas in several regions where both steam and water are produced at temperatures over 200°C. In addition, geothermal resources are used in many district and process heating schemes.

A number of test boreholes have been drilled to explore the potential in the UK; the deepest was at Larne (Northern Ireland), and others were at Cleethorpes, Southampton and Marchwood (near Southampton). The results have been generally disappointing and estimates of the available resource in the UK are small. The ownership of the borehole at Southampton, funded by Government, was transferred to Southampton City Council. The Southampton district heating scheme represents the only exploited geothermal heat in the UK.

Municipal Wastes, General Industrial Wastes and Landfill Gas

The disposal of wastes produced by households, industry and commerce can pose a number of environmental problems. However, these wastes can also be used as a source of energy. Energy can be recovered by combustion of the waste as collected (mass-burn incineration) or after processing to reclaim recyclable components (such as metals and glass) or by utilising the methane-rich gas produced by biological processes that occur when waste is landfilled (landfill gas collection and utilisation) or sewage sludges are treated.

SELCHIP: South East London Combined Heat and Power waste combustion plant.

The potential for recovering energy from waste, and the mix of technologies that will be employed in the future, is inevitably determined by trends in waste disposal practice. At present, over 90% of all household and commercial waste produced in the UK is landfilled. Organic wastes decay in landfills in the absence of oxygen producing a mixture of primarily methane and carbon dioxide known as landfill gas. Emissions of landfill gas, if not controlled, can have serious environmental consequences. Techniques for the collection and use of landfill gas have been developed over the past 10 years, and are now in commercial operation. Current trends in environmental policy and legislation favour the development of larger engineered landfills, remote from centres of waste production and the closure of smaller sites. There will also be tighter controls on environmental standards. Waste disposal costs are therefore expected to increase.

BFI's Packington landfill site unloading refuse.

Incineration is well-established overseas as a means of treating a wide range of wastes. The capital and operating costs of incineration plant are high and in the UK the availability of low cost landfill has proved a major barrier to development. There are also concerns about emissions from incinerators, particularly of heavy metals and dioxins, which have resulted in the introduction of strict emission regulations. Techniques for producing and using fuels generated from refuse have been under development for some years. Work in the UK has largely focused on the production of a densified refuse derived fuel, for use in industrial boilers. Work is now concentrating on energy production with materials recovery in centralised resource recovery facilities.

General industrial waste produced by commerce and industry is primarily handled by the private waste disposal industry. These wastes generally have a high calorific value and provide greater opportunities to segregate materials for recycling, with residues available for use as fuel. This material could be used in smaller, less costly incineration plant.

Specialised Industrial Wastes

Industry produces a diverse range of specialised wastes which require disposal. Many of these have some inherent energy value and, if burnt as a fuel, the total resource would have an energy content equivalent to some 3.5 million tonnes of coal per year.

Tyres, one type of specialised industrial waste.

Very little use is presently made of the resource. This is generally because wastes arise in small quantities from specific operations in discrete locations, and are therefore regarded by industry as a nuisance to be dealt with at the lowest cost or with the minimum of inconvenience. In the UK this has usually meant direct disposal to landfill through a private waste contractor. Where used, incineration has been employed as an adjunct to disposal, principally as a means of reducing disposal costs rather than as a means of producing useful energy.

Agricultural and Forestry Wastes

Agricultural and forestry wastes fall into two main groups, dry combustible wastes such as forestry wastes and straw, and wet wastes such as "green" agricultural crop wastes (i.e. root vegetable tops) and farm slurry. The former group of biofuels are utilised using thermal processes to give heat directly (via combustion), or converted into a second fuel either gaseous (via gasification) or liquid (via pyrolyis). The latter group of biofuels are best utilised via anaerobic digestion to produce methane ("biogas").

Farm waste digestion plant.

Currently, very little use is made of these materials as sources of energy, despite the fact that frequently there is a cost associated with their clean disposal. For example, surplus straw now has to be ploughed into the soil and animal slurries must be contained to prevent water course adulteration. Conversion of these wastes into fuels can generally be accommodated within existing agricultural and forestry practice. Thus in future these wastes may be considered as additional income earners.

The use of fuels derived from agricultural and forestry residues could create markets which energy crops might then supply at a later date. In Denmark today, there is 50 MW$_e$ of straw burning plants and in the USA wood fuelled power stations total approximately 6,000 MW$_e$.

Energy Crops

Crops which may be grown to produce energy in the UK range from food crops grown for energy purposes to woody biomass. From these sources solid, liquid or gaseous biofuels may be derived. Many methods for the conversion of biomass are available, reflecting the diversity of the resource. The drier, lignin-rich materials (e.g. wood) are best suited to combustion, gasification or pyrolysis conversion processes. Wetter biomass can be converted through anaerobic digestion to a methane-rich biogas fuel. Other fermentation techniques produce liquid fuels such as ethanol.

Well-developed systems of varying scale are available to provide direct heat or electricity from the crop. Conventionally, electricity is produced by

Harvesting willow.

burning wood in a boiler to generate steam that is fed to a turbine. More advanced technologies involving gasification are ready for demonstration. They should allow electricity production at higher efficiency and lead to significant reductions in costs.

Trials have indicated that arable coppice – particularly willow and poplar – appear the most promising option for the UK as these crops can provide high yields and a favourable overall energy balance. The first large scale commercial demonstration projects are now underway. These will allow current estimates of yields and costs to be confirmed and demonstrate that fuel from this source can be produced commercially. Significant improvements in yields and reduction in production and harvesting costs can be expected as experience is gained.

There are now significant opportunities for the production of energy crops as a result of the reform of the Common Agricultural Policy, a central element of which is the reduction of food overproduction. New measures will result in some farm land becoming potentially available for non-food crops, including energy crops. The benefit of this approach is that such land will remain productive and not fall derelict. This will go some way to maintaining farm incomes and rural economies.

World interest in biomass is growing rapidly with major programmes underway in Scandinavia, the USA and Europe. Significant energy crop enterprises already exist in Brazil and Sweden with development programmes underway in Scandinavia, Europe and North America. Over 6,000 ha of short rotation coppice have been established in Sweden alone.

Advanced Conversion Technologies

Enniskillen gasification plant.

There are a number of so called advanced conversion technologies that are being considered as alternatives to conventional steam raising plant for converting biofuels. Principal amongst these are pyrolysis, gasification and liquefaction. These thermochemical processes produce solid, liquid and gaseous intermediates from biofuels, which can be used to produce electricity and/or heat, or be upgraded to directly substitute for fossil fuels. The intermediates can be used in an engine to produce power, avoiding the use of a steam cycle. This gives a very significant increase in conversion efficiency with comparable capital and operating costs at the scales relevant to biofuels (less than 50 MW_e). In this way, the economic viability of electricity production from energy crops, forestry wastes and straw is significantly improved.

These processes promise to be inherently less polluting than conventional incineration. By reducing the costs of pollution abatement and offering a more secure disposal route in the light of ever more stringent pollution legislation, the incorporation of these technologies in energy from waste schemes is likely to increase. For sewage sludge, gasification may prove to be the best option for both disposal and power generation as conventional incineration pushed to the limits is barely autothermic.

In the Scandinavian paper industry over 100 MW_{th} of biomass gasification plant has been in operation for many years. Major advanced biomass gasification projects for power generation are now underway in Sweden, Finland and Hawaii. In the UK the planning of a 5 MW_e plant is well advanced.

Summary of the Current Status of Renewable Energy Technologies

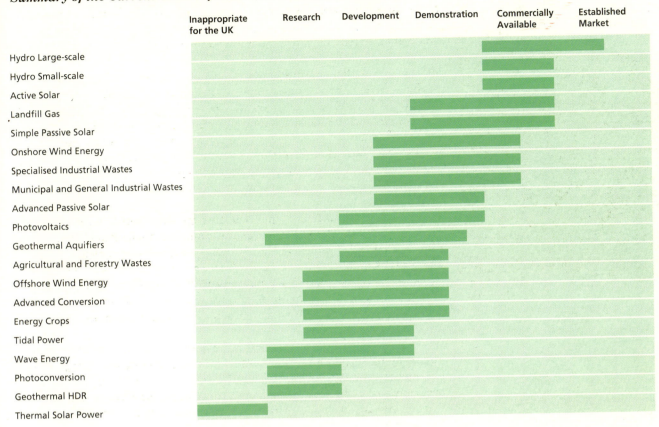

Figure 2. Current status of technologies in the UK.

Key Characteristics of the Renewable Energy Technologies

As a class of technologies the renewables have a number of important features that distinguish them from conventional energy technologies. Although each of these features does not necessarily apply to all renewables, certain common characteristics are worth discussion.

Some technologies such as wind, hydro and geothermal aquifers have been used for many centuries. However, in their modern form, most renewables are relatively new technologies for energy provision and have yet to be widely accepted as technically mature and commercially viable.

Many renewable energy technologies are available in a wide range of capacities. Wind turbines and hydro schemes are available from a few kilowatts up to multi-megawatts as single plant. Most other technologies can be considered modular in that they can be readily aggregated to make a larger system. This range in available capacities allows renewables to be tailored to meet the characteristics of particular resources and markets. For example, a wind farm can be designed to be of a sufficient capacity to meet the specific energy requirements of a community.

Renewables tend to have a high initial investment cost and low operating and maintenance costs. This is a natural consequence of technologies utilising free resources with a high conversion cost. The characteristic "up-front" cost of renewables means that, when compared to conventional technologies with flatter cost schedules, renewables are more sensitive to the discount rate used in their economic appraisal. An economic appraisal with a high discount rate will give the greatest weight to the costs and receipts which occur early on in the life of a project and renewable energy projects therefore appear relatively more attractive when assessed with low discount rates.

Technology	Scale	Investment Cost £/kW	O&M Costs £/kW/year	Fuel Costs £/kW/year	Load Factor (%)	Construction Time (years)	Lifetime (years)
Onshore Wind Energy	1 kW to 3 MW	800 to 1,200	12 to 18	0	25 to 40	1	15 to 25
Hydro Power Large-scale	over 5 MW	1,000 to 3,000	2.5 to 7.5	0	35	4	100+ civils 15 to 50 M&E
Hydro Power Small-scale	5 kW to 5 MW	600 to 3,000	12 to 60	0	15 to 95	1	25 to 40
Tidal Power	30 MW to 9 GW	1,200 to 1,500	5 to 10	0	20 to 25	3 to 10	100+ civils 30 to 40 M&E
Wave*	1MW to 2 GW	1,500 to 3,000	40 to 90	0	7 to 25+ device dependent	1 to 10	25 to 35
Photovoltaics (new buildings)	0.001 to 0.5 MW	1,500 to 1,700	unknown	0	12 to 15	0.2 to 1	25
Active Solar (domestic hot water)	0.5 to 5 kW$_{th}$	500 to 1,000	0 to 3	0	not applicable	0.05	unknown
Passive Solar Design	500 to 30,000 kWh$_{th/year}$	£0 to £50/GJ	0	0	0	1	60
Geothermal HDR*	3.3 MW	12,000	300	0	90	5	20
Geothermal Aquifers (notional UK scheme)	10 MW$_{th}$	550 to 650	15 to 30	0	30	2	15 to 20
Waste Combustion	5 to 40 MW	2,500 to 5,000	120 to 150	-400 to 0	85	2 to 3	20
Landfill Gas	0.6 to 3 MW	790	130	0	88	1	10
Agricultural Wastes: Anaerobic Digestion Combustion	0 to 1 MW 5 to 40 MW	4,000 to 5,000 1,200 to 1,800	50 to 100 60 to 120	with credits -100 to 0 0 to 300	95 85	1 to 2 1 to 2	20 20 to 25
Forestry Residues: Combustion	5 to 30 MW	800 to 1,500	60 to 120	100 to 330	85	1 to 2	20 to 25
Energy Crops: Combustion Gasification	5 to 30 MW 0.1 to 30 MW	800 to 1,500 600 to 1,500	60 to 120 50 to 100	150 to 250 120 to 160	85 85	1 to 2 1 to 2	20 to 25 20

Table 1. Typical characteristics of current renewable energy technologies. * Assumes successful completion of RD&D.

Definition of terms.

Scale: The range of scheme size in watts installed electrical power (W$_e$) or watts of heat (W$_{th}$).
Cost/kW: The cost in £ per kilowatt (1992) of installed capacity.
O&M Costs: Annual costs in £ per kilowatt (1992) of installed capacity – excluding rates and insurance.
Fuel Costs: Annual costs in £ per kilowatt (1992) of installed capacity.
Load Factor: The ratio of the actual power sent out in an average year to that which would be delivered if the plant were able to run at full capacity for the whole year. This measure takes account of the availability of both the resource or fuel and the generating plant.

Some of the renewables are only able to provide power intermittently. The output from sources such as wind, solar and wave is variable in both timing and power, as they are dependent upon the prevailing weather conditions and thus unpredictable for more than a short time ahead. Tidal power schemes also provide power intermittently but their output is entirely predictable. The biofuels are exceptions to this rule as they utilise an organic resource which can be stored, like a fossil fuel, for use when required. The variable and energy limited nature of renewables is sometimes cited as an impediment to their integration within a conventional supply system. However studies have shown that intermittent electrical generation from renewables should be able to provide up to 20% of a grid system's demand with only a modest reduction in the value to the system of the renewable energy. Greater penetrations are technically feasible.

Notwithstanding these particular features, there are many similarities between renewables and conventional technologies and renewables are capable of being incorporated within a modern power supply system with only minor adjustment to conventional practices.

Some of the key characteristics of renewable energy technologies are presented in table 1. For most technologies the cost data is for plant currently available. For many technologies these costs can be expected to reduce with further development and deployment. The cost of energy is discussed in Section 2.

1.2. The Marketplace

Where the renewables are currently being exploited and why

Most renewables are not currently competitive, under current accounting systems, with conventional generation technologies for the bulk provision of electricity for centralised despatch via a high voltage integrated grid system. As integrated grids and central dispatch are the modus operandi for the electricity systems of developed countries the opportunities for renewable energy deployment within the developed world, without subsidy, are limited at present. The large scale introduction of renewables awaits the anticipated reduction in renewable energy generation costs and the adoption of accounting systems that address the environmental costs of energy generation. There are, however, certain niche markets in developed countries where renewable energy technologies are currently competitive. These are:

- where there is an exceptional resource allowing economic generation; for example, certain wind sites, hydro sites and refuse deposits;

- where centrally dispatched electricity is unavailable and the cost of conventional forms of generation are prohibitively expensive; for example in remote areas and for island communities;

- where the renewable scheme is part of a larger or existing project and only part of the project costs are attributed to the renewable energy scheme; for example, a geothermal aquifer assisted district heating scheme, passive solar design in new buildings, landfill gas control, and some energy from waste schemes;

- where consumers wish to make a commitment to renewable energy; for example, domestic active solar.

United Kingdom

The development of renewable energy technologies in the UK has been spearheaded by the Department of Trade and Industry's Renewable Energy Programme, managed by ETSU. Initiated in 1974, in the wake of the world oil crisis, the programme has spent over £330 M to date (1993 money). The aim of the Department's programme has been to stimulate the development of the renewable energy technologies to the fullest practical extent where they have prospects of being economically attractive and environmentally acceptable. This aim is being pursued by means of a collaborative research, development and demonstration programme with industry and by the establishment of an institutional framework which will ensure that renewables can compete in the market on equal terms with other energy sources.

The renewable energy market has been given a tremendous stimulus by the introduction of the Non-Fossil Fuel Obligation (NFFO) for renewable energy. The Electricity Act 1989 gave the Secretary of State powers to make Orders obliging the Regional Electricity Companies (RECs) to contract for specified minimum amounts of non-fossil fuel sourced electricity generating capacity. The RECs are compensated for the extra cost by means of a levy on electricity consumers.

So far, two renewable NFFO Orders have been made. In pursuit of these two orders the RECs signed nearly 200 contracts, with a total capacity of over 600 MW. Under the first Order, in 1990, the RECs signed 75 contracts for a total of 152 MW DNC (Declared Net Capacity). Under the second Order in 1991 they signed 122 contracts for a total of 472 MW DNC. A summary of the first and second rounds is given in table 2.

Technology/Energy Source	Number of Contracts	Size of Contracts MW	Contracted Capacity MW	Contract Price p/kWh
First Round 1990				
Landfill Gas	25	0.18 to 4.8	35.5	
Sewage Gas	7	0.17 to 3	6.5	
Municipal and General Industrial Waste	4	1.65 to 27	40.6	
Hydro Electricity	26	0.04 to 3.8	11.8	
Wind	9	0.039 to 3.9	12.2	
Other	4	0.1 to 20	45.5	
First Round Totals	75		152.1 MW (DNC)	
Second Round 1991				
Landfill Gas	28	0.4 to 5.0	48.5	5.7
Sewage Gas	19	0.3 to 7.8	26.9	5.9
Municipal and General Industrial Waste	10	4.1 to 103	271.5	6.55
Hydro Electricity	12	0.02 to 0.6	10.9	6.0
Wind	49	0.1 to 7.4	84.4	11.0
Other	4	1.0 to 12.5	30.2	5.9
Second Round Totals	122		472.2 MW (DNC)	
First and Second Round Totals	197		624.3 MW (DNC)	

Table 2. Summary of first and second round NFFO contracts in England and Wales.

The Declared Net Capacity (DNC) for intermittent renewables plant is very broadly defined as the equivalent capacity of base-load plant that would produce the same average annual energy output. For wind energy the DNC is taken as being 43% of the installed capacity. For the other technologies shown here the DNC and the installed capacity are identical.

The European Commission first approved the NFFO/Levy arrangements only up to 1998 and contracts under the first two renewables orders expire

Technology	MWe
Onshore Wind Energy	124
Hydro Power – Large-scale	1,360
Hydro Power – Small-scale	78
Shoreline Wave	0.1
Photovoltaics	0.2
Landfill Gas	77
Municipal Waste Combustion	52
Sewage Sludge Digestion	104
Farm Wastes	25
TOTAL	1,820

Table 3a. Deployment of renewables for electricity generation in the UK in September 1993.

Technology	Thousands tonnes of oil equivalent
Active Solar Heating	7.5
Geothermal Aquifers	0.6
Landfill Gas	29.4
Sewage Sludge Digestion	40.6
Wood Combustion	163
Straw Combustion	67.1
Municipal Waste Combustion	41.1
Other	75.7
TOTAL	425

Table 3b. Renewables used to generate heat in the UK, 1992.

on that date. However the Commission has extended this deadline in relation to renewables allowing future orders to go beyond 1998. These prices should not be taken as indicative of future contract prices under NFFO type arrangements.

The successful application of the NFFO has produced an active market for renewable energy in the UK. There are now many companies with a commercial interest in renewable energy be it as owners, manufacturers or providers of the many different services required during the installation and operation of renewable energy plant. Many of the companies involved with renewable energy in the UK, as a result of the NFFO, are UK owned or have a UK base for their operations.

The deployment of renewables in the UK is presented in tables 3a and 3b. These figures do not take account of plant contracted under the 1990 and 1991 NFFO orders that have recently been commissioned and other plant still awaiting commissioning. The figures are from a survey of available data and may contain minor errors.

World-wide

Today, approximately 20% of the world's electricity production comes from renewable energy sources. Existing installations are mainly hydroelectric, but there is approximately 10 GW of electricity generation from wood and over 3 GW of wind turbines installed with over 1 GW in Europe. Many current studies are predicting significant future contributions from the renewables. For example; the World Energy Council estimates that by 2020 the renewables contribution to total world energy supply could be almost twice the 1990 level and a recent study for the United Nations considers that by the middle of next century renewable sources of energy could account for three-fifths of the world's electricity market and two-fifths of the world market for fuels used directly. The potential world market for renewable energy technologies is therefore very large. By way of an example, if an additional 1% of the world's electricity supply were to come from renewable energy it would require an investment in new plant of approximately £25bn.

Recognition of the role of renewables in assisting the European Union to meet recent environmental commitments suggests that major developments in the renewable energy market are likely to take place within Europe in the near future. The European Union's ALTENER (Alternative Energy) Programme is designed to increase the development of renewable energy sources in the Union and increase trade in products, equipment and services within and outside the Union. The European Union's objective is to double the renewables contribution to primary energy consumption from 4% in 1991 to 8% in 2005.

The size of the market in the industrialised world outside the USA and the European Union is difficult to quantify, but there is presently considerable activity in Japan, Sweden, Canada, Israel and Australia. The renewables would seem to have prospects for deployment within Eastern Europe, where there is a shortage of power together with pollution problems from out-dated conventional technology. There is however little internal money available to finance activities and foreign aid will be required to initiate any major programme.

The range in size, modularity, the relative ease with which most renewables can be constructed and commissioned and the ubiquitous nature of renewable resources makes them attractive as stand-alone energy providers within both the industrialised world and the developing world. This may be a very large market for renewables in the future.

	Fossil	Nuclear	Renewables
All OECD	60%	24%	16%
North America – OECD	64%	20%	16%
Pacific – OECD	67%	20%	13%
European Union	58%	34%	8%
Belgium	39%	60%	0%
Denmark	98%	0%	2%
France	14%	74%	13%
Germany	70%	28%	3%
Greece	91%	0%	9%
Ireland	95%	0%	5%
Italy	79%	0%	21%
Luxembourg	88%	0%	12%
Netherlands	95%	4%	0%
Portugal	70%	0%	30%
Spain	46%	36%	18%
UK	77%	22%	1%

Table 4. Percentage of electricity generated from fossil fuels, nuclear and renewables in 1991: source OECD 1993.

1.3. Opportunities and Constraints

The development of a self-sustaining market for renewable energy in the UK depends upon a number of factors. Renewable energy technologies will not be freely deployed until the market is satisfied that:

- there is a suitable resource available for commercial exploitation;

- they are at least as economically attractive as competing technologies (including subsidies);

- they have achieved satisfactory technical standards (safety, reliability, availability, maintainability, durability etc.);

- their entrance to the market is not significantly affected by institutional and regulatory difficulties.

This section examines the opportunities and constraints facing the renewable energy technologies today which are classified here under the headings of technical, institutional and environmental. The question of economic competitiveness is dealt with in detail in Section 2.

Technical Opportunities and Constraints

The prospects for most renewable energy technologies could be significantly improved with further development work in a number of technical areas. There are opportunities to increase the available resource, decrease costs, and to improve the manufacturability, reliability, durability and market compatibility of the technologies. Improvements in any one of these areas may have a significant effect on the prospects for a technology.

Resource Access and Utilisation

Energy production from renewables could be increased in two ways. Firstly, by accessing more resource. Secondly, by utilising the available resource more efficiently. The larger the resource base and the more efficient the technologies the greater the prospects for their deployment. There are many opportunities for developments in these areas of which some examples are presented below.

Increasing the resource base
- The biomass crop resource could be increased by the development of hardier, higher yielding plants.

- The development of larger wind turbines would allow the extraction of more energy from the winds over the land on which they are sited.

- Access to renewable energy resources could be improved by mapping and monitoring their location and providing the information to potential developers.

Improving utilisation efficiency
- The potential for energy recovery from landfilled wastes could be increased by improved engineering of landfill sites and optimising landfill management practices so that microbes are able to digest waste more effectively.

- The development of advanced conversion technologies – gasification and pyrolysis – which are more efficient than conventional steam raising plant, would increase the energy yield from biofuels.

- There are a number of improvements that could be made to wind turbine blade design which could lead to increased energy output for a given machine height.

- The further development of photovoltaic materials should result in a steady improvement in the direct conversion efficiency of sunlight to electricity.

Reliability, Durability
Most energy users are used to, expect and require a secure, low maintenance energy supply. Operators will demand systems with a high degree of reliability and low maintenance requirements; particularly if the renewables are to be used for decentralised applications as a substitute for grid power. For many renewables there are opportunities for improvements in these areas.

Market Compatibility
To-date most development effort has gone into the engineering of practical renewable energy systems. Having developed working technologies there is scope for the development of products that are compatible with the requirements of particular markets.

- Improvements to the design of renewables – fewer components, modular/repetitive design – could reduce the time and expense of installation and repair.

- The stand-alone operation of intermittent renewable energy systems is likely to require some measure of energy storage. There are opportunities for the development of improved storage systems such as advanced high energy batteries, very high speed flywheels, compressed air storage and hydrogen conversion and storage techniques.

Manufacturability

There is considerable scope for reducing the manufacturing costs of most renewable energy technologies by improved design, focusing particularly on production engineering issues. To date the low demand for technologies such as photovoltaics, wind turbines and biofuels plant has not allowed manufacturers to design machines for long production runs which would reduce unit cost. The move towards volume production must occur if these technologies are to become competitive and this will require significant levels of investment.

The Role of RD&D

Current technical constraints hindering the deployment of renewable energy technologies may be removed and opportunities for improving the potential of renewables realised by the judicious use of research, development and demonstration effort. The potential for improvement through RD&D is summarised in table 5. This categorisation takes no account of a technology's prospects for economic deployment and therefore makes no judgement of the value of particular RD&D. Clearly, it would be inappropriate to undertake RD&D activities for technologies having little or no prospect of deployment.

Technology	Resource Access	Utilisation Efficiency	Reliability/ Durability	Market Compatibility	Manufactur- ability
Onshore Wind Energy	Medium	Medium	High	Medium	High
Offshore Wind Energy	Low	Medium	High	Medium	High
Hydro Power (large & small)	Low	Low	Low	Low	Low
Tidal Power	Low	Low	Low	Low	Medium
Wave Energy	Low	High	High	High	Medium
Geothermal HDR	High	Medium	High	High	Medium
Geothermal Aquifers	Low	Low	Low	Medium	Medium
Photovoltaics	Low	High	Medium	High	High
Photoconversion	Low	High	High	High	Medium
Active Solar	Low	Low	Low	Low	Medium
Passive Solar Design	Low	Low	Low	High	Medium
Municipal and General Industrial Wastes *	Low	Low	Low	Low	Medium
Landfill Gas	Low	Medium	Medium	Low	Low
Specialised Industrial Wastes *	Medium	Medium	Medium	Medium	Medium
Agricultural and Forestry Wastes *	Medium	Medium	Medium	Medium	Medium
Energy Crops *	Medium	High	Medium	Medium	Medium

High ▮ (orange)   Medium ▮ (green)   Low ▯ (light)

Table 5. Potential for improvements through RD&D.
* These technologies could utilise advanced conversion technologies.

Institutional Opportunities and Constraints

Most renewables are relatively new energy technologies and as such are often contrasted with the "conventional" technologies. The structure, modus operandi and prejudices of the energy industry, allied service industries, Government and legislative bodies have yet to adjust to the "unconventional" needs of these newcomers. Widespread experience of renewables will be required before their needs can be accommodated as readily as those of the conventional technologies.

Financing Difficulties

The viability of many renewable energy schemes is critically dependent upon the ability of the developer to secure financing on acceptable terms. Renewable energy schemes are highly capital intensive and require developers to raise finance for their schemes either by drawing on internal reserves (where available), or externally through corporate loans or via project specific financial structures which frequently require complex contractual arrangements.

The discount rates used for assessing and financing renewables are usually higher than those used for the assessment of established technologies. This is because the deployment of renewable energy technologies is generally perceived to bear more risk. This perception is largely due to investors lack of awareness of renewable energy technologies and the absence of an industry with a proven track record.

Successful demonstration projects can help to reduce the perceived risk associated with renewables and thereby increase their deployment. Reducing the cost of finance, increasing its availability and easing contractual constraints would also help. Investigating the extent to which obtaining finance is likely to represent a barrier to renewable energy and identifying the major hurdles to be overcome is the first step towards addressing this problem.

Regulatory Constraints

The entry of renewables into the market has been hindered by the existing regulatory framework which was not designed with the needs of renewable energy technologies in mind. Those wishing to deploy renewable energy technologies have encountered difficulties with the interpretation and appropriate application of regulations concerned with such matters as environmental protection, pollution control, waste management and disposal, conservation areas, planning consent and water abstraction. However, regulation could be used positively to stimulate investment. For example, tightening building regulations can stimulate investment in both energy efficiency and passive solar design; waste disposal regulations can determine the disposal option chosen.

A regulatory framework which would ensure that renewables can compete in the market on equal terms with other energy sources – "a level playing field" – would support the introduction of renewables. Early identification of potential regulatory problems would help in the design of an accommodating framework.

Planning Requirements and Perception of Environmental Impacts

Generally renewable energy technologies are perceived as "green" and beneficial to the global environment. However, on a regional or local scale they are not always viewed as such and this leads to the questioning of the desirability of renewables in certain locations. This is underlined by the growth of the NIMBY "not in my backyard" and BANANA "build absolutely nothing anywhere near anything" syndromes. It is understandable that planning authorities are reticent about the approval of such new schemes and that, at this early stage, the statutory and non-statutory bodies concerned with planning applications may object. This is reflected in delays in the planning process which may be compounded by the need to go to public inquiry. The costs of the planning process can be a significant burden on a small project, particularly if a full Environmental Impact Assessment is required.

Planning policy guidelines have now been produced by the Department of the Environment and the Welsh Office. The Scottish Office has published its own National Planning Policy Guidance for public consultation. These guidelines cover renewables generally and a specific annex has been produced on wind. Annexes on other renewable energy technologies will follow. In addition, the DTI has initiated a number of studies in England and Wales with both county and district councils to inform them about the renewable energy technologies, the resources and the planning and environmental issues and to assist them in the incorporation of renewable energy policies within their development plans.

Lack of Market Incentives

A major obstacle to the deployment of renewable energy is the lack of both an established market and an equipment supply industry. Many potential developers still regard the renewable energy technologies as unproven and therefore risky investments so that even after demonstrations have been carried out it is difficult for the technologies to become established. This presents something of a "chicken and egg" problem: until a market is established there will be a lack of confidence in the technologies; until there is confidence in the technologies a market will not be established.

These difficulties have largely been resolved for the electricity generating renewables by the creation of an initial, temporary, protected market under the NFFO legislation discussed earlier. An analogous scheme for heat generating renewables would be beneficial.

Industry Infrastructure

The renewable energy industry includes equipment manufacturers and installers, operators and maintainers and other interested parties such as trade associations and professional institutions. The effectiveness of the industrial infrastructure will determine the rate of deployment of renewable energy technologies.

When industries are growing steadily, free market forces should provide the necessary work force. However, when the market is new and demand is created rapidly it can be expected that, in the short term, demand for skilled and experienced personnel will outstrip supply. A lack of skilled personnel – equipment manufacturers, installers and operators, planners and financiers – could constrain the uptake of renewable energy technologies. At best a skills shortage could slow market development. More seriously, it could result in poorly implemented projects which could seriously affect the reputation of renewable energy.

The development of the necessary skills resource, and with it the development of the market, could be assisted by the provision of suitable training and guidance.

Lack of Awareness of Renewables

Lack of awareness of the opportunities afforded by renewables will constrain their uptake. The knowledge which has been gained through RD&D programmes needs to be transferred to industry, commerce and the general public. This should aid the acceptance and understanding of the renewables and accelerate their deployment. This dissemination process should be targeted towards key market sectors.

Technology	Regulatory/Planning	Financial	Infrastructural
Onshore Wind Energy	High	Medium	High
Offshore Wind Energy	Medium	High	Medium
Hydro Power (large & small)	High	Medium	Medium
Tidal Power	High	High	High
Wave Energy	Low	High	High
Geothermal HDR	Medium	High	High
Geothermal Aquifers	Medium	High	High
Photovoltaics	Medium	High	Medium
Photoconversion	Low	Low	Low
Active Solar	Medium	Medium	Low
Passive Solar Design	High	Medium	Low
Municipal and General Industrial Wastes *	High	Medium	Low
Landfill Gas	Medium	Medium	Low
Specialised Industrial Wastes *	High	Medium	Low
Agricultural and Forestry Wastes *	Medium	Medium	Low
Energy Crops *	Medium	High	Medium

High [ ]   Medium [ ]   Low [ ]

Table 6. Significance of institutional factors.
* These technologies could utilise advanced conversion technologies.

The significance of the institutional constraints for the individual technologies is summarised in table 6.

Environmental Opportunities and Constraints

Concern about the environment and our impact upon it is greater now than it has ever been. Of the many environmental issues debated today the environmental and health impacts of different energy systems, particularly those associated with the production of electricity, are amongst the most significant.

Anxieties about the appropriate choice of electricity generating systems, from the viewpoint of the environment, are highlighted by the current debate on the effects of pollution, climate change, acidification of forests and lakes, depletion of the ozone layer, depletion of scarce resources and the potential for severe accidents such as the one at Chernobyl.

Renewable energy technologies are seen by many as part of a solution to these problems. In their operation they produce little or no net emissions of polluting gases, they don't deplete scarce, irreplaceable resources and they are relatively safe in operation and maintenance. Of course, the deployment of renewable energy technologies would have environmental impacts but they tend to be of a different nature to those associated with fossil and nuclear technologies. A convenient way of considering their environmental impacts, and contrasting these impacts with those of conventional technologies, is to consider where the impacts occur geographically.

The environmental impacts of energy technologies may occur at a local level (within a few miles), a regional level (county, national and near international), or at a global level. In general the environmental impacts of renewables are greatest at the local level with many of the technologies having no direct regional or global environmental impacts. The exceptions to this rule are the biofuels technologies where the burning of a fuel and the subsequent release of combustion products would have environmental impacts far beyond the immediate locality. However, as the following discussion shows, the use of biofuels can have considerable advantages.

Local Effects

The local impacts of energy technologies are varied. Clearly, all energy technologies have a physical presence and their location can be an extremely controversial matter. Smaller plants, which the renewables tend to be, offer more opportunity for a discreet location but with small plants more sites would be needed to satisfy a given energy demand. Technologies that are integrated into buildings, such as photovoltaics, active solar and passive solar design are the least intrusive.

As well as having a visual impact most technologies have a land requirement. As non-fuel burning renewables utilise resources which are diffuse and of a low energy density they tend to have plant which is larger per unit capacity than conventional technologies. However, because of the generally benign nature of their operation there are less restrictions on public access to their sites. So, for example, tidal barrages can be considered as bridges and most of the land on which wind turbines are sited can retain its original use.

There is always some noise associated with the generation of high voltage electricity but most plant can be placed in sound-proof housing. An exception is wind turbines where the swishing noise of the moving blades is sometimes remarked upon. However with good design and careful siting there is no reason why wind turbines should be a nuisance.

The waste and biomass combustion technologies, in common with fossil fuel burning technologies, have local environmental impacts that include emissions to atmosphere, fuel transportation and handling activities and the disposal of ash and other wastes. There may also be requirements for clean water and the disposal of effluents. However, it must be remembered that the generic characteristic of wastes is that they are an undesired by-product of other activities. The environmental impact of using these residues as fuel must therefore be assessed within the context of other waste disposal or minimisation options.

Renewable energy technologies which do not use a fuel only exhibit local environmental impacts in their operation. Only the biofuels technologies have impacts that go further afield. The local environmental impacts of renewable energy technologies are not of a severe or dangerous nature, generally being ones of "nuisance".

Regional Effects

Fuel burning plants produce conversion products that can be transported and deposited over considerable distances creating a regional or international environmental impact. The most notorious problem is that of sulphur dioxide which, in the air, can damage plant life and, once deposited, can cause acidification of streams, rivers, lakes and soil. Present problems with sulphur dioxide arise from the burning of coal and oil which both contain sulphur up to 6% and 3% respectively. Municipal and industrial wastes also contain sulphur at generally lower levels (typically less than 1%) but contain generally higher levels of chlorine which may be emitted as hydrochloric acid. However, stringent emissions regulations apply to waste technologies so that acid gas emissions are only a fraction of those that occur from coal and oil burning plant, even those fitted with modern flue gas desulphurisation equipment. Future European regulations for waste burning plant are likely to reduce acid gas, particulate and other emissions even further.

Global Effects

For many years the atmospheric concentrations of carbon dioxide, nitrogen oxides, methane and other radiatively active gases have been increasing due to human activities. These gases play an important role in climate control and evidence that their increase may be bringing about climate change has caused considerable concern. Together the gases form a transparent blanket around the earth admitting short wave length solar radiation but retaining the longer wave length radiation from the earth. This blanket of insulating gases allows the earth to maintain a mean temperature of around 15°C rather than approximately -10°C that would occur if the gases were absent. This is known as the "greenhouse effect".

The most significant greenhouse gas is carbon dioxide because, although on a molecular basis it is not the most active, human activities generate far more carbon dioxide than any other greenhouse gas. The huge quantities of carbon dioxide that are produced annually from the combustion of hydro-carbon fuels mean that containment of significant quantities would be so difficult and expensive that it is not practicable. Reducing the amount of carbon dioxide produced is widely seen as an important way of mitigating climate change and as at least one third of Europe's annual production of carbon dioxide is from power generation, the substitution of renewables for fossil plant is seen as an important part of any carbon dioxide abatement strategy.

Most renewable energy technologies produce no carbon dioxide or other greenhouse gases in operation. Renewable energy technologies that use biomass crops and crop wastes as fuel produce carbon dioxide, but since the carbon dioxide was absorbed and fixed by photosynthesis while the crop was growing the carbon is effectively cycled with no net addition to atmospheric carbon dioxide. Of course, cycling of the carbon requires that there is no permanent reduction in the crop base; land used for energy crops being replanted. Renewable energy technologies that utilise municipal and industrial wastes also produce carbon dioxide. As the wastes have to be disposed of, the greenhouse gases produced by alternative disposal routes need to be considered. The most likely disposal routes are landfill or incineration. If wastes are landfilled, their natural anaerobic decomposition will produce methane, a particularly potent greenhouse gas. From the point of view of global warming it is beneficial to burn this methane, because although this will produce carbon dioxide, the contribution to global warming will be less than if the methane were allowed to enter the atmosphere. In addition, if the energy is utilised it avoids generation elsewhere. The direct combustion of wastes produces more energy per unit of carbon dioxide than landfilling and produces more energy and carbon dioxide per unit of waste as plastics and other materials that would not have decomposed in a landfill are burnt.

Technology	Local	Regional	Global
Renewables			
Onshore Wind Energy	Noise, Visual Intrusion, Electromagnetic Interference	None	None
Offshore Wind Energy	Impeded Navigation and fishing rights	None	None
Hydro Power	Visual Intrusion, Ecological Impact	None	None
Tidal Power	Visual Intrusion, Ecological Impact	Ecological Impact	None
Wave Energy	Impeded Navigation and fishing rights	None	None
Geothermal HDR	Noise, Visual Intrusion, Water Requirements, Radon release	None	None
Geothermal Aquifers	Visual Intrusion	None	None
Photovoltaics	Visual Intrusion	None	None
Active Solar	Visual Intrusion	None	None
Passive Solar Design	None	None	None
Photoconversion	Visual Intrusion	None	None
Municipal and General Industrial Wastes *	Particulate and Toxic Emissions, Visual Intrusion, Fuel Transportation	Particulate and Toxic Emissions, Waste Disposal	None
Landfill Gas	Particulate and Toxic Emissions, Visual Intrusion	Particulate and Toxic Emissions	None
Specialised Industrial Wastes *	Particulate and Toxic Emissions, Visual Intrusion, Fuel Transportation	Particulate and Toxic Emissions, Waste Disposal	None
Agricultural and Forestry Wastes *	Particulate Emissions, Visual Intrusion, Fuel Transportation	Particulate Emissions	None
Energy Crops *	Particulate Emissions, Visual Intrusion, Fuel Transportation	Particulate Emissions	None
Conventional			
Coal	Particulate and Toxic Emissions, Visual Intrusion, Fuel Transportation	Particulate and Toxic Emissions, Waste Disposal	Carbon dioxide release
Oil	Particulate and Toxic Emissions, Visual Intrusion, Fuel Transportation	Particulate and Toxic Emissions	Carbon dioxide release
Gas	Particulate and Toxic Emissions, Visual Intrusion	Particulate and Toxic Emissions	Carbon dioxide release
Nuclear	Visual Intrusion, Accidental leakage of radioactive material	Waste Disposal Possible major accident	Possible major accident

Table 7. Summary of the environmental impacts of renewable and conventional energy technologies.
* These technologies could utilise advanced conversion technologies.

Technology	
Tidal Power	Flood protection Potentially additional amenities such as a road crossing, marina and watersport developments
Hydro Power (large & small)	Flood protection Oxygenation of water
Passive Solar Design	Improved living conditions
Municipal and General Industrial Wastes	Waste Disposal
Landfill Gas	Waste Disposal Energy production from methane that would otherwise escape to the atmosphere contributing to global warming
Specialised Industrial Wastes	Waste Disposal
Agricultural and Forestry Wastes	Waste Disposal
Energy Crops	Land management Greater variety of flora and fauna

Table 8. Summary of the additional environmental benefits of renewable energy technologies.

2. Potential and Prospects for Renewable Energy

2.1. Size of Resources and Current Uptake

In this work, two terms are used to describe resource sizes: the Accessible Resource and the Maximum Practicable Resource. These need some explanation to allow correct interpretation.

The Accessible Resource represents the resource which would be available for exploitation by a mature technology after only primary constraints are considered. For example; for wind power National Parks and physical constraints, such as housing, roads and lakes are excluded from the calculation of the resource size. For most technologies this measure of resource is still large and its full exploitation unlikely to be acceptable as it would result in power plant in every available location. While the Accessible Resource may therefore indicate a considerable theoretical potential for renewable energy, it is not a realistic measure of the actual contribution which may be made in future. The Accessible Resource includes existing schemes where appropriate. An explanation of the derivation of the Accessible Resource for each technology is given within the individual technology modules in Annex 1.

Estimates of the Accessible Resource for the renewable energy technologies in the UK are presented in the following graph.

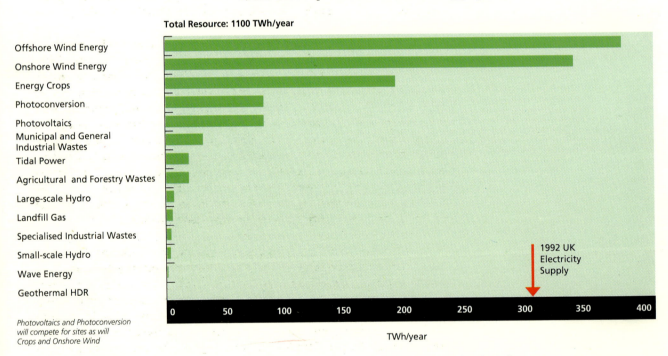

Photovoltaics and Photoconversion will compete for sites as will Crops and Onshore Wind

Figure 3. Accessible Resource for electricity producing renewable energy technologies at a cost of 10p/kWh or less (1992) 8% discount rate.

To assess the contribution that renewables might make in the real world a more realistic measure of their resource, taking account of additional constraints upon their deployment, is required. This measure is here called the Maximum Practicable Resource. In deriving the estimates of the Maximum Practicable Resource that are presented in this document an examination of the constraints on the deployment of each technology, and how they might change with time, was undertaken. Many of these constraints – regulatory, sociological, environmental – are not susceptible to objective scientific assessment and subjective judgements were often required in order to allow the derivation of figures for the Maximum Practicable Resource. These judgements were informed by studies undertaken as part of the earlier programme and existing experience of deployment which in most cases is extremely limited. Moreover for each technology this work assesses the UK as a whole without detailed analyses of the complex system integration and operational issues at a regional level, but the Programme's regional assessments have suggested that these issues could significantly constrain the Practicable Resource. The resource estimates presented here are therefore for the "Maximum" Practicable Resource.

The results of this assessment of the Maximum Practicable Resource and the cost of its exploitation are summarised in the supply curves; figures 4-7. These curves present the estimated resource that could be supplied by each of the technologies as a function of its cost of economic exploitation (1992 money) for the years 2005 and 2025 assessed with two different discount rates of 8% and 15%. These estimates include existing schemes where appropriate. An explanation of the derivation of the Maximum Practicable Resource for each technology is given within the individual technology modules in Annex 1. These curves need to be treated with care. In particular the various assumptions and caveats given in the technology modules, which vary from technology to technology, should be noted.

As technologies are developed it is reasonable to expect that, with time, their costs will come down and their ability to effectively exploit their resource increase. These are the principal reasons why the Maximum Practicable Resource is estimated to be cheaper and greater in 2025 than in 2005.

The approximate saving in carbon dioxide emissions that would occur from using a given amount of generation from renewable energy technologies is also shown. These estimates assume that each unit of renewable energy generation effectively displaces a unit of conventional generation and its concomitant emissions. As it is not possible to say what specific conventional generation would be displaced by new renewable energy generation – it might be coal, gas, nuclear or other depending on the location and the level of uptake of renewables – these estimates assume that the carbon dioxide emissions displaced are equivalent to those from an average unit of plant on the current (1990) UK electricity system. The emission savings can then be established by subtracting the net emissions produced by the renewable energy plant during operation from the displaced emissions.

Most renewables plant does not produce carbon dioxide during operation. The exceptions are those using biofuels. Of these, the biomass technologies – energy crops and agricultural and forestry wastes – produce no net carbon dioxide emissions during operation as the carbon is effectively being recycled. Wastes going to landfill generate methane gas which, if not used for heat or power generation, would need to be flared. It is therefore assumed that the use of landfill gas for energy production also results in no net carbon dioxide emissions. The situation is more complicated for the two remaining biofuels – municipal wastes and industrial wastes. A detailed consideration of the nature of these wastes and alternative disposal routes is required to derive the net emissions during operation. For simplicity it is assumed here that municipal wastes and industrial wastes produce no net emissions during operation. However, within their technology modules in Annex 1 the detailed analysis is presented.

In the future it is anticipated that gas turbines, which produce less carbon dioxide per unit of energy delivered, will take a much greater share of the market. If this occurs, the emissions savings are likely to be less than shown

Supply curves for electricity producing renewable energy technologies.

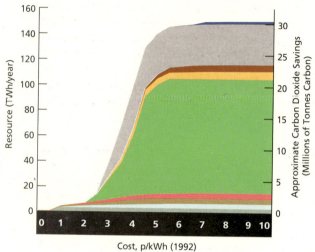

Figure 4. Maximum Practicable Resource, 2005; 8% discount rate.

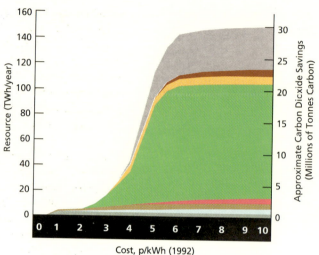

Figure 5. Maximum Practicable Resource, 2005; 15% discount rate.

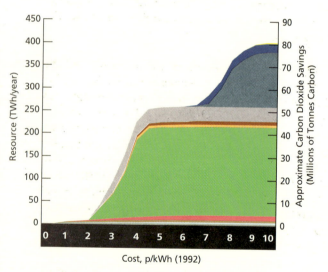

Figure 6. Maximum Practicable Resource, 2025; 8% discount rate.

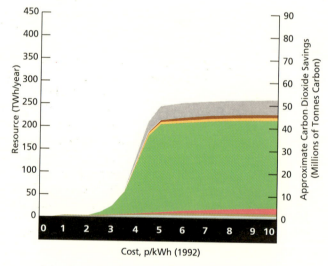

Figure 7. Maximum Practicable Resource, 2025; 15% discount rate.

here. However, in the short to medium term it is likely that renewables will displace coal plant with consequently higher emission savings.

Within this assessment it has been assumed that biofuels resources will be used exclusively for electricity generation. This assumption greatly simplifies the analysis and is considered reasonable as electricity generation is likely to be the main market for these resources. However some of the resources will be used for heat production, this may be together with electricity in combined heat and power plant (CHP) or in simpler plant designed for heat production alone. In either case the efficiency of heat producing plant will be higher than that of plant which only produce electricity and, if it is the preferred option, the costs will be lower.

Three renewable resources which produce only heat have been considered in this assessment: geothermal aquifers, passive solar design and active solar heating. The supply curves for the UK geothermal aquifer resource are presented in its technology module in Annex 1. A supply curve can not be readily constructed for passive solar design. This is because the costs associated with passive solar technology are unique to the type of installations being considered – domestic or commercial, new-build or refurbishment – and are often negligible. However, data describing the passive solar design resources and costs are presented within its technology module. The active solar resources are also difficult to present as supply curves. Active solar is one of very few technologies where the domestic consumer is the main "developer", and the economic criteria which are applied to bulk supply technologies are inappropriate. Simple data describing the active solar resources and costs are presented within its technology module.

Supply curves are a useful way of describing renewable energy resources and their costs. However, when using them, one should be aware that most of the renewable energy technologies are still being developed and that estimates of their future costs, performance and available resource are uncertain, particularly for those at an early stage of assessment, such as energy crops, where the uncertainties are greater. The supply curves are not, therefore, forecasts of what will occur but realistic assessments of what could be supplied if things go well for renewables and appropriate RD&D is undertaken. As the supply curves are the input data to the energy systems modelling exercise, which addresses energy demand as well as energy supply, these points should be borne in mind when considering the results presented in the following section.

2.2. Energy Systems Modelling

The contribution that any technology may make in the future will be determined principally by the availability of a commercial technology and of an exploitable resource, the economic competitiveness of the technology with other forms of generation (including those already supplying the system) and the demand for energy. It is clear therefore that supply curves present only part of the story. In order to consider the role that particular renewables might play in a future UK energy market, it is necessary to consider not only the renewables market but also the evolution of the whole UK energy market, as developments here will determine the willingness of suppliers to enter and develop the renewables sector. This is notoriously difficult to predict and a range of views of the future were therefore considered and their effect on the development of renewables assessed.

The methodology adopted for the assessment utilised a scenario approach combined with energy system analysis using the MARKAL (MARKet ALlocation) model developed by the International Energy Agency. An important aspect of the study was to gauge how robust the estimated potential contributions of the technologies were to uncertainties regarding future international prices for primary energy and UK demands. This was achieved by running the model with prices and demands developed in scenarios which conceptualised very different economic and social backgrounds for the evolution of the UK and the international energy sectors. These scenarios, all of which were given equal weighting in the assessment, had the following themes.

i. High Oil Price (HOP). This scenario envisaged a future in which oil and gas prices rise steeply to the highest levels considered sustainable, but coal prices remain low.

ii. Composite (CSS). A scenario capturing a range of conventional thinking on the future development of prices and demands at the outset of the study in 1990.

iii. Low Oil Price (LOP). A future in which oil and gas prices remain at their present low levels.

iv. Heightened Environmental Concern A (HEC-a). A future in which society and economic management are strongly influenced by environmental concerns. This is manifest by a high carbon tax aimed at reducing carbon dioxide emissions coupled with other measures to encourage initiatives such as recycling and greater use of public transport. Also the scenario envisaged that there would be a moratorium on the construction of nuclear power plant and an early rundown of existing stations.

v. Heightened Environmental Concern B (HEC-b). Essentially the same scenario as HEC-A except that additional nuclear development is permitted.

vi. Shifting Sands (SS). Strictly, not a scenario in its own right but a sensitivity test imposed on the CSS scenario to investigate the effect which oil price "shocks" would have on the potential contribution of technologies.

It should be noted that these scenarios are not predictions of future UK energy prices and demands. The scenarios are "what if" tools which provide a reference framework for assessing the prospects of the energy technologies and the robustness of these prospects to future uncertainties.

The analysis of the size and timing of the potential contributions for each scenario was made using the MARKAL model. The model is based on a linear programming system for minimising an "objective function", which in this case was the discounted cost of the overall energy system. The model yields a least cost solution for the energy system which gives the mix of technologies adopted to meet externally defined demands. For this study, the period 1995 to 2025 was considered, providing an assessment of the energy mix at 5 year intervals with results in terms of deployment against time for each technology considered. Given the necessary database, the model can also quantify the environmental emissions associated with the energy system and can have emissions constraints imposed upon it in order to study the most cost-effective technology responses to measures aimed at reducing environmental burdens. (A fuller discussion of the application of the energy systems model is presented in the documentation for the Appraisal of UK Energy RDD&D – volume 8).

The cost and performance data for each energy technology that might potentially be deployed in the UK, both conventional and renewable, were input to the model. The data used for the renewable energy technologies is consistent with the supply curves for the Maximum Practicable Resource shown in the previous section. Where the supply curves for a technology may be significantly affected by future environmental policy, regulations etc., a second set of supply curves for the Maximum Practicable Resource, applicable to the Heightened Environmental Concern, or "greener", scenarios was provided. This approach was adopted for five technologies, landfill gas, municipal solid wastes, specialised industrial wastes, photovoltaics and passive solar design. These supply curves, appropriate for a more environmentally sensitive future than that if current trends continue, are presented in the respective technology modules within Annex 1.

Generally the model operates with a single uniform discount rate applying to all technologies. For this assessment two such rates (8% and 15%) were investigated. These values were chosen as being reasonably representative of the rates required by major companies when considering main stream investments. In addition a "survey" case was investigated in which a 10% rate was assigned to main stream investments whilst a 25% rate was applied to non-core energy investments (e.g. energy efficiency) for a range of reasons such as availability of capital, information barriers, hidden costs etc.

The assessment of six scenarios against three different discounting regimes resulted in the consideration of 18 possible futures. The total uptake of all electricity producing renewable energy technologies under these futures is shown in figures 8 – 10.

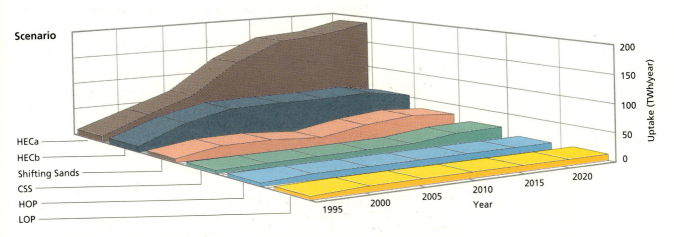

Figure 8. Renewable energy electricity generation under various future scenarios; 8% discount rate.

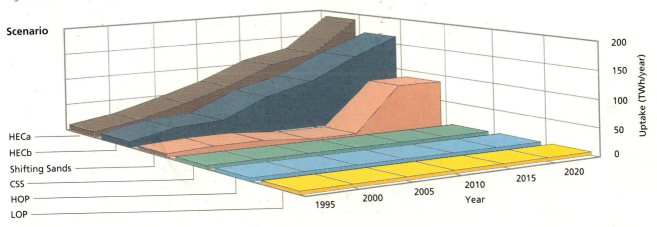

Figure 9. Renewable energy electricity generation under various future scenarios; 15% discount rate.

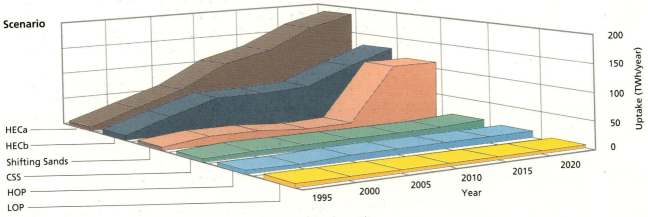

Figure 10. Renewable energy electricity generation under various future scenarios; survey rate.

Estimates of future renewable energy generation need to be considered within the context of UK total electricity supply and how this changes with both time and scenario. Table 9 presents estimates of total electricity generation, for all technologies including both centralised and decentralised plant. Alongside these estimates the percentage of this supplied by renewable energy technologies is shown.

	Scenario	1990	1995	2000	2005	2010	2015	2020	2025
Total Electricity Generation (TWh/year)	CSS	290	280	270	280	290	300	310	320
	LOP	290	280	280	290	290	300	300	320
	HOP	290	280	280	280	280	290	300	310
	HECa	290	260	250	250	250	260	270	280
	HECb	290	260	250	250	250	260	270	280
	Shifting Sands	290	280	280	280	290	300	330	350
Renewables as a % of Total Electricity Generation	CSS	2%	2%	3%	3%	4%	4%	4%	5%
	LOP	2%	2%	2%	2%	2%	2%	1%	1%
	HOP	2%	2%	2%	3%	4%	4%	4%	4%
	HECa	2%	4%	15%	27%	43%	50%	63%	68%
	HECb	2%	4%	15%	26%	26%	27%	38%	46%
	Shifting Sands	2%	2%	6%	6%	4%	4%	34%	33%

Table 9: UK electricity generation under different scenarios and the contribution from renewable energy; Survey discount rate.

The proportion of the total amount of renewable energy generation supplied by any particular technology varies with scenario. The following graphs show the make-up of renewable energy electricity generation for the six scenarios assessed with the 8%, 15% and Survey discount rates for the years 2005 and 2025.

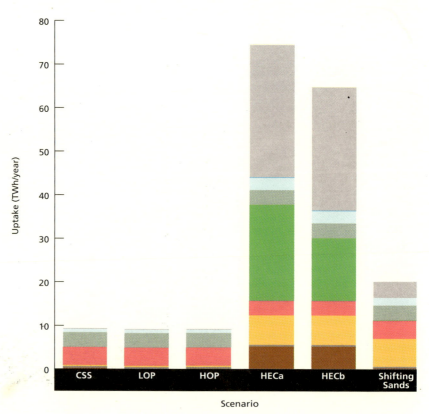

Figure 11. Renewable energy electricity generation under various future scenarios; 2005, 8% discount rate.

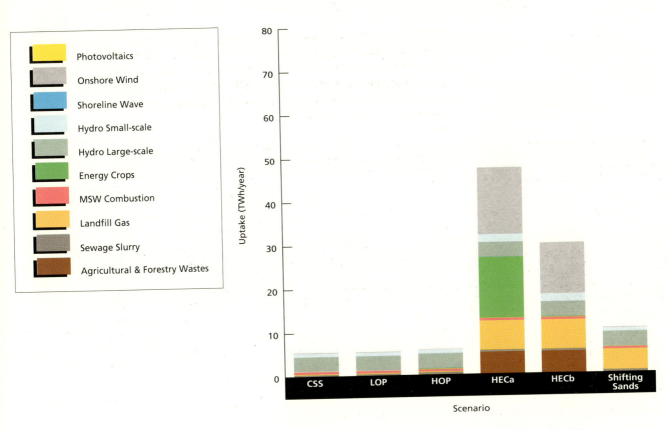

Legend:
- Photovoltaics
- Onshore Wind
- Shoreline Wave
- Hydro Small-scale
- Hydro Large-scale
- Energy Crops
- MSW Combustion
- Landfill Gas
- Sewage Slurry
- Agricultural & Forestry Wastes

Figure 12. Renewable energy electricity generation under various future scenarios; 2005, 15% discount rate.

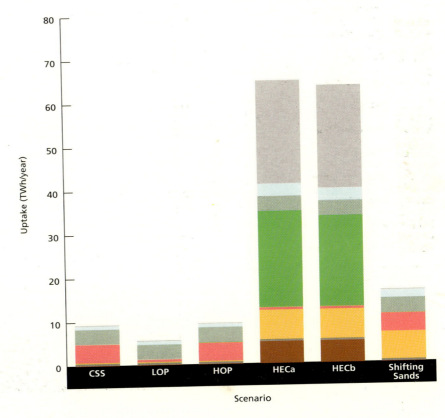

Figure 13. Renewable energy electricity generation under various future scenarios; 2005, Survey discount rate.

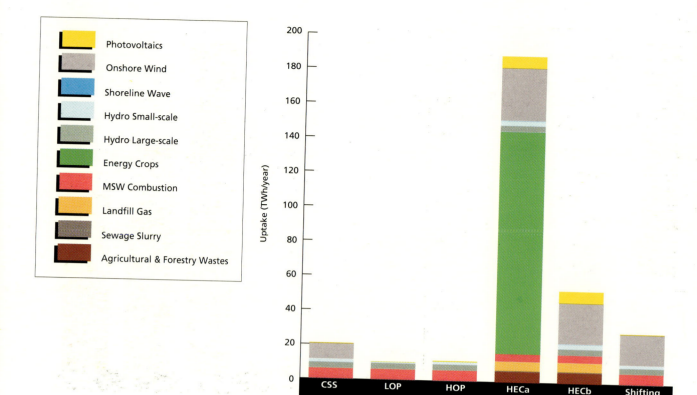

Figure 14. Renewable energy electricity generation under various future scenarios; 2025, 8% discount rate.

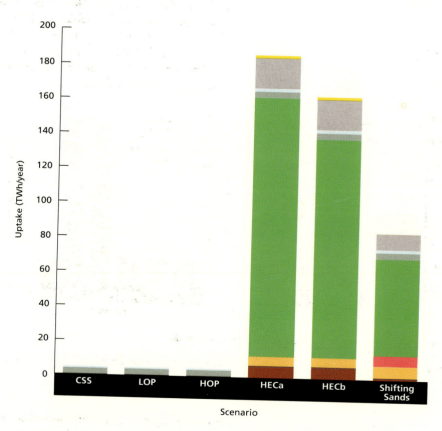

Figure 15. Renewable energy electricity generation under various future scenarios; 2025, 15% discount rate.

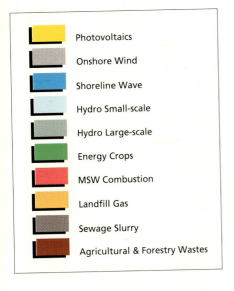

- ▮ Photovoltaics
- ▮ Onshore Wind
- ▮ Shoreline Wave
- ▮ Hydro Small-scale
- ▮ Hydro Large-scale
- ▮ Energy Crops
- ▮ MSW Combustion
- ▮ Landfill Gas
- ▮ Sewage Slurry
- ▮ Agricultural & Forestry Wastes

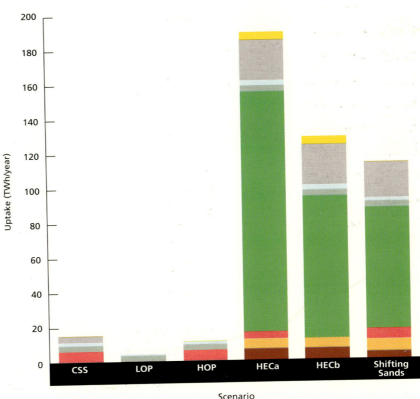

Figure 16. Renewable energy electricity generation under various future scenarios; 2025, Survey discount rate.

Figures 17 – 19 show the make-up of renewable energy heat generation for the six scenarios assessed with the 8%, 15% and Survey discount rates for the years 2005 and 2025. Geothermal aquifers are not shown as the model chose not to deploy the technology in any time period under any scenario. The estimates for passive solar design only include new-build and a minimal estimate of refurbishment. For the year 2005 the choice of discount rate has no effect on the results and for 2025 the results are the same for the 15% and Survey discount rates.

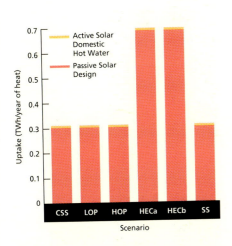

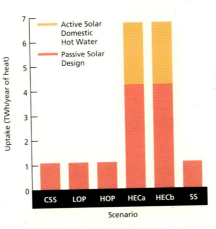

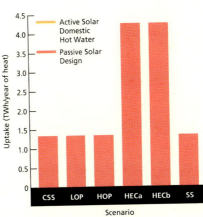

Renewable energy heat production under various future scenarios.

Figure 17. 2005, All discount rates.

Figure 18. 2025, 8% discount rate.

Figure 19. 2025, 15% and Survey discount rates.

The estimates for PSD only include new-build and a minimal estimate of refurbishment.

To some degree, the performance of a technology in the HEC scenarios may be regarded as a measure of its environmental impact. In these scenarios, a heavy penalty was placed on carbon emitting technologies through the carbon tax assumed to be imposed, and tightened constraints were imposed on emissions of other environmental burdens such as sulphur dioxide and nitrogen oxides. However, the scenarios assume a particular mix of penalties/constraints so they do not offer a test of a technology's response to restriction of any individual burden. In addition, it may be felt that the penalties/constraints selected for the scenarios are unrealistically tough, so that a less demanding test may be of interest.

An additional form of environmental impact assessment was therefore undertaken. For two scenario/discount rate cases, gradual reductions were imposed in system emissions of carbon dioxide, sulphur dioxide and nitrogen oxides. The response of each individual technology was assessed in terms of any significant increase – or decrease – in its contribution under the relevant emission constraint.

The two scenarios chosen were the HOP and CSS scenarios. Neither includes any assumptions on increasing environmental regulation in the scenario definition, and both offer a very different broad picture of the pattern of energy supply in terms of fuel preference. The survey discount rate regime was chosen in each case. No-one would claim that the cost optimised picture presented by the MARKAL energy system model represents a projection of how the world will turn out in practice, but of all the runs undertaken, those utilising the survey discount rate regime are likely to come the closest to modelling a realistic pattern of technology take-up.

For both these cases, constrained runs of the MARKAL model were undertaken as follows:

i. total system emissions of carbon dioxide were reduced from 1990 levels at 5% per decade to the end of the scenario period;

ii. total system emissions of carbon dioxide were reduced from 1990 levels at 10% per decade to the end of the scenario period;

iii. total system emissions of sulphur dioxide were reduced from 1990 levels at 40% per decade to the year 2010, and maintained at that level thereafter;

iv. total system emissions of nitrogen oxides were reduced from 1990 levels at 40% per decade to the year 2010, and maintained at that level thereafter.

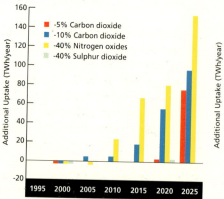

Figure 20. Additional contributions to the Composite scenario resulting from the imposition of environmental constraints for electricity generating technologies.

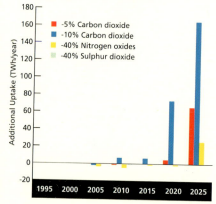

Figure 21. Additional contributions to the High Oil Price scenario resulting from the imposition of environmental constraints for electricity generating technologies.

2.3 Discussion of Results

Examination of the preceding graphs clearly shows that the renewable energy technologies do not react to the different scenarios in a uniform manner. It is also clear that their response to particular scenarios varies with discount regime. This is not surprising as there are marked differences in the renewables cost, performance and resource characteristics and their variation with time. For these reasons, an assessment of the future prospects for the renewables is better undertaken by examining how the individual technologies respond to the scenarios rather than by attempting to draw conclusions from their collective potential. The analysis of the modelling results for the individual technologies is presented within the individual technology modules.

Notwithstanding the previous comments, there are some general conclusions that can be drawn from the modelling exercise.

The renewables contribute most under the Heightened Environmental Concern scenarios where the constraints upon the amount of carbon dioxide the system can emit causes the cost of energy to rise and certain renewables to become competitive. At all three discount rates, by 2005, renewables are providing more than 30 TWh/year of electricity (10% of current supply) under the Heightened Environmental Concern scenarios and by 2025 they are supplying between 55 and 190 TWh/year. In later years much of this generation would be expected to come from energy crops. As for all scenarios, these estimates assume successful completion of appropriate RDD&D, the resolution of system integration issues and the success of the Government's programme of market enablement measures. In the particular case of energy crops, these estimates assume an estimated projection of land surplus to requirements which is heavily dependent upon Common Agricultural Policy reforms.

Under the Composite, Low Oil Price and High Oil Price scenarios the scenario criteria can be met using conventional fossil technology and there is little new renewable energy build. The renewables uptake that is shown is predominantly that from the existing large-scale hydro and the first two rounds of the NFFO.

The Shifting Sands scenario, which models fluctuations in the price of fossil fuels based on the Composite scenario fuel prices, results in a contribution from the renewables that lies between that from the carbon dioxide constrained scenarios and the non-carbon dioxide constrained scenarios and varies markedly with time as the fluctuations in the price of fuels causes the renewables to become periodically competitive.

The choice of discount rate has a significant effect on the contribution from renewables. As expected, the higher discount rates favour conventional technologies over renewables. There is one interesting exception however. Under the Heightened Environmental Concern scenario in which new nuclear build is allowed, HEC(b), there is more renewables deployment with a 15% discount rate than with an 8% discount rate. This occurs because nuclear plant, with their large up-front costs and long construction times, are even more adversely affected by high discount rates than renewables.

All main stream energy investments, including renewables, were assessed using the same discount rate for each of the three discount regimes considered. In practice, developers of renewables will expect a rate of return that adequately reflects the risks associated with a project, over a time period that reflects commercial financing realities. This may be considerably shorter than the lifetime of the project. Consequently, the discount rates used to assess renewable energy technologies in the commercial market may vary considerably from project to project and the investment costs will differ from those presented in this assessment if project risk contingency is built into commercial estimates. The element of risk will be greater for the less established technologies, unique projects and technologies with innate uncertainties – for example, geothermal technologies where the quality of the heat resource may not be properly established prior to considerable investment. As, in a sense, a risk-neutral assessment is presented here, it should be noted that the more "risky" technologies have fared well in comparison with those which are tried and tested.

The imposition of additional environmental constraints upon the Composite and High Oil Price scenarios generally leads to a significant, and increasing, additional renewables uptake after 2010. In earlier periods the additional constraints have little effect.

Under both scenarios, the requirement for an additional 5% per decade decrease in carbon dioxide emissions can be met without a significant increase in renewables uptake until 2025 when in all cases an additional renewables uptake of at least 65 TWh/year by 2025 is achieved. The requirement for an additional 10% per decade decrease in carbon dioxide emissions results in additional renewable uptake in both scenarios and all time periods after 2010. Under the High Oil Price scenario and 10% carbon dioxide constraint the additional uptake is as high as 165 TWh/year in 2025. It is worth noting that the 10% carbon dioxide constraint generally has more than double the effect of the 5% carbon dioxide constraint on renewables uptake.

The additional nitrogen oxides constraint results in a significant and increasing uptake of renewables after 2010 under the CSS scenario but has little effect under the HOP scenario. This is a reflection of the availability of a low nitrogen oxides producing fossil fuel technology – integrated gasification combined cycle – together with a low coal price under the HOP scenario.

The additional sulphur dioxide constraint had no effect on the uptake of renewables as the constraint can be met by fuel switching between conventional technologies; principally a switch to gas-fuelled generation.

3. Conclusions

1. There are significant, technically exploitable, renewable energy resources available in the UK.

2. In the UK some renewable energy technologies are currently commercial in niche markets while others will require further RDD&D if they are to become competitive. However, some are unlikely to be commercialised in the foreseeable future.

3. If the costs of conventional generation remain at current levels there is unlikely to be more than a modest increase in the contribution from renewable energy to future UK energy provision over the next two to three decades.

4. If the costs of conventional generation rise, or if the external environmental costs of generation are taken into account, the economics of many renewables will compare favourably with those of conventional forms of generation; indeed they could make a substantial competitive contribution over the next two to three decades.

5. If concern over the environment heightens requiring a reduction in carbon dioxide emissions further than the current commitment, renewable energy is likely to be a necessary and cost-effective part of a solution.

6. Prospects for the UK deployment of individual renewable energy technologies between now and 2025 are varied.

   - Within the short term, those with the best prospects include: onshore wind, small-scale hydro, landfill gas, waste combustion and passive solar design.

   - Energy crops, utilising advanced conversion technology, have a substantial potential and offer prospects of commercial deployment in the medium term.

   - Active solar, decentralised photovoltaics and photoconversion offer prospects for modest commercial deployment but require further assessment before this can be confirmed.

   - Tidal power, offshore wave, offshore wind, centralised photovoltaics, geothermal technologies and new large-scale hydro are unlikely to be commercially deployed in the foreseeable future. Shoreline wave has a very limited potential for commercial deployment.

7. To fulfil their potential, most renewable energy technologies would require further development work in a number of technical areas. There are opportunities to increase the available resource, decrease costs, and to improve the manufacturability, reliability, durability and market compatibility of the technologies. Improvements in any one of these areas may have a significant effect on the prospects for a technology.

8. The commercial exploitation of most renewable energy sources depends on relatively new technologies. The structure and practices of the traditional energy industry, allied service industries, Government and regulatory bodies will need to adjust to the "unconventional" needs of these newcomers.

9. If renewable energy technologies are to be readily deployed in the UK, an infrastructure appropriate to the particular needs of renewable energy will need to develop. In particular, a skills base of experienced equipment manufacturers, installers and operators, planners and financiers – together with financial, planning and regulatory systems responsive to the needs of renewables – will be required.

10. From a global viewpoint the renewable energy technologies are environmentally benign and can provide a major contribution to sustainable development. In common with all forms of energy provision the renewable energy technologies have environmental impacts but unlike conventional generation technologies these are generally of a local and reversible nature. Planning procedures sensitive to the needs of local communities are required if the renewable energy technologies are to fulfil their potential to improve the global environment.

11. The world-wide market for renewable energy products and services is potentially enormous. There is already a large overseas market for decentralised applications and the market for centralised generation at home and abroad is growing steadily. If this potential is to be realised it will require the development of a renewable energy industry far greater than that currently existing. The disparity between the potential world market and the current world industry offers the opportunity for the development of a significant UK industry for domestic and export sales.

Annex 1: Technology Modules

Wind Power

I. Technology Review

A. Technology Status

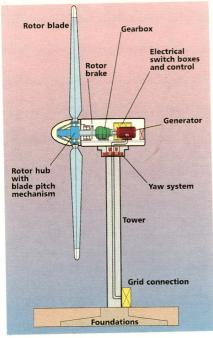

Figure 1: Cutaway of horizontal axis wind turbine.

Wind is a very complex resource. It is intermittent and strongly influenced by geography and topography (terrain effects). There is a non-linear (cubic) relationship between instantaneous wind speed and available power.

Wind power has been harnessed by man for over 2,000 years through the construction of windmills. The use of wind as a renewable energy source involves the conversion of power contained in masses of moving air into rotating shaft power. The conversion process utilises aerodynamic forces (lift and/or drag) to produce a net positive turning moment on a shaft, resulting in the production of mechanical power which can then be converted to electrical power.

There are two basic design configurations for wind turbine generators (WTGs) depending on the axis of rotation of the drive shaft horizontal axis (HA) machines (like a traditional windmill) and vertical axis (VA) machines. Horizontal axis designs are at a more advanced stage of development and are more cost-effective at the current state of commercial development. For machines of greater than 1MW capacity, vertical axis designs may be cheaper than horizontal axis ones and are claimed to offer a scope for lower electricity generation costs in the future. Figure 1 shows the component parts of a typical horizontal axis wind turbine.

Wind turbines are being developed over a range of power outputs from kilowatt to multimegawatt units and are available commercially up to around 500 kW. Large scale generation of electricity requires a number of machines (typically around 20) to be grouped together for economy and ease of operation. The machines are usually spaced 5 to 10 rotor diameters apart to reduce interaction effects, which impair their performance, to an acceptable level. As a result, a wind farm of about 20 machines usually extends over some 3 to 4 square kilometres of land although the actual machines occupy only about 1% of this area and the remaining land can be farmed as usual. Figure 2 shows a typical wind farm.

Figure 2: The Wind Farm at Cemmaes (courtesy of National Wind Power Ltd).

The net energy output of a typical 400 kW machine operating in a wind farm would be around 1,000 MWh/year on a site with an annual mean wind speed (AMWS) of 7.5 metres per second (m/s) at a height 25 m above ground level (AGL) and 1,400 MWh/year on a site with an AMWS of 9.0 m/s at 25 m.

The technology is well established with over 20,000 grid connected machines (with an average rating of about 100 kW) in operation world wide. Apart from the need to demonstrate adequate fatigue life and reliability there is no doubt about the technical feasibility of wind power. Existing design philosophies aim at machine lifetimes of around 25 years; this can probably be achieved although experience of operating modern wind turbines is limited to around 10 years. The reliability of wind turbines has increased steadily during the past decade as better designs have evolved; modern turbines are now operating with a machine availability of over 95%.

Decommissioning wind turbines at the end of their design life presents few difficulties as it only entails the removal of scrap material and cabling. The concrete bases are a few cubic metres in size and can be removed, though they are usually buried. There is no residual waste or land contamination.

There is the potential to develop wind farms using offshore locations. The technology for offshore deployment is the same as that for onshore but the harsher climate and relative inaccessibility place more stringent requirements on the initial design and subsequent reliability. Additionally, because the foundations of a machine become a much greater proportion of the total project cost, multi-megawatt rated machines will need to be deployed to optimise the cost of energy. Such machines would require a significant development and demonstration programme stretching over decades rather than years.

B. Market Status

Onshore wind energy is one of the more promising renewable energy sources for electricity generation world-wide. The Government has encouraged the commercial exploitation of wind energy in England and Wales through the Non-Fossil Fuel Obligation (NFFO) and 58 wind projects totalling 225 MW were included in the 1st and 2nd tranches of the Obligation. By the end of 1993, 19 wind farms with a total capacity of over 100 MW were operating or being installed in England and Wales.

In continental Europe and the USA uptake of the technology is further advanced than in the UK. In the USA, capacity installed since 1980 is in excess of 2,000 MW. The rate of installation declined following the withdrawal of tax concessions; however, the industry is now recovering and the rate of installation is increasing again as the cost of the energy from wind systems falls. In Europe, over 300 MW has been installed in Denmark (over 50% owned by private individuals) and over 150 MW in both the Netherlands and Germany. Many countries have targets of several hundreds of megawatts by the year 2000, or soon after. Consequently there should be a buoyant market and considerable export potential for those countries manufacturing the most competitive machines. The current status of the technology is summarised in table 1.

Figure 3: Estimated annual mean wind speeds for mainland Great Britain, given in metres per second at 25 m above local ground level.

UK installed capacity	>100 MW
European Union installed capacity	>1,000 MW
World installed capacity	>3,000 MW
UK industry	One established and several potential manufacturers
UK business	Several developers and consultants

Table 1: Status of wind energy technology (early 1993).

C. Resource

Onshore

Wind Speeds and Land Area.
Weather patterns in the northern hemisphere result in the UK having one of the best wind resources in Europe. Mean wind speeds increase with the elevation of the land so the areas in the UK with the greatest potential lie in the hilly regions of the country. Wind speeds also increase with height above ground level. Annual mean wind speeds have been calculated for the UK at a resolution of 1 sq.km using computer modelling techniques which include topographical effects. These have been normalised to long term data obtained for 51 Meteorological Stations. Calculations indicate that there are around 12,500 sq.km (5% of UK land area) in England and Wales and over 37,500 sq.km (15% of UK land area) in Scotland having an AMWS of greater than 7.5 m/s at 25 m AGL. Figure 3 shows some results of these calculations.

Land Usage Limitations
There is a limit on the availability of land for the siting of wind turbines. Physical limitations include features such as towns, villages, lakes and woods. Other constraints include designated areas such as National Parks, regional parks, areas of outstanding natural beauty, national scenic areas, sites of special scientific interest, local and national nature reserves, and environmentally sensitive areas. There may be scope for siting some turbines in these areas but each case would merit special consideration.

Siting Limitations
The number of turbines which could be erected on nominally constraint-free land is limited because turbines can only be sited some distance from houses and not too close to roads, isolated patches of woodland and power lines.

Accessible Onshore Resource
Onshore resource estimates have been made using annual mean wind speed data and a specific set of land usage limitations. The effects of siting limitations have been estimated through a sampling technique using a land classification system. An existing, validated system of 32 land classes has been used. The methodology adopted took the following steps.

- Land areas excluded from consideration for siting by defined physical and institutional constraints were identified.

- For each land class, each square kilometre of the remaining land area was classified as a function of annual mean wind speed, at 0.5 m/s intervals.

- For each land class, the average number of wind turbines which could be sited on a square kilometre of the non-excluded land areas was estimated using 12 sample squares (each 1 km by 1 km) and taking into account defined local siting limitations. Turbines of 33 m diameter (300 kW rated power) were used, being typical of leading commercial machines at the time. Each machine had to be located 330 m or more away from houses or from each other and a minimum distance of 150 m from roads, isolated patches of woodland and power lines.

- For each land class, the energy capture within each wind speed band was estimated using the appropriate number of turbines per square kilometre and the power performance curve for a single, typical 300 kW machine.

- Correction factors were applied to the energy capture of single machines to account for specified wind farm operating losses.

The total over all land classes provided an estimate of the Accessible Resource as a function of AMWS; the total over all wind speed bands gave a total Accessible Resource in the UK of about 340 TWh/year, reflecting the windy nature of the British climate.

The Accessible Resources for specific areas in the UK are: England & Wales 120 TWh/year, Scotland 190 TWh/year and Northern Ireland 33 TWh/year. For comparison, the annual electricity usage in the UK is around 300 TWh/year.

Offshore

Sea Area
The wind resource offshore is potentially much greater than onshore and is limited only by practicable working water depths and use of maritime areas for other activities. Moving from onshore to offshore need not require major changes in the technology but additional technical problems arising because of the more hostile environment will need to be overcome. In the long term, offshore installations could increase the contribution from wind power substantially.

Offshore Wind Speeds
Average offshore wind speeds are higher than those onshore. There is little reliable data on which to make large scale estimates, but computer modelling methods have been developed which predict average wind speeds for areas offshore measuring 10 km by 10 km.

Offshore Siting Constraints
Siting is considered practicable in water depths down to 30 m and the main impediments to exploitation are alternative usage of offshore areas (e.g. fishing, shipping lanes, military test grounds). The number of available sites has been obtained from an early offshore study carried out for the then Department of Energy in the mid-1980s. Areas were identified around the UK coast where it was considered feasible to construct large wind turbines. Siting constraints imposed by other offshore activities were identified and the relevant areas excluded from the analysis.

Accessible Offshore Resource
Taking workable water depth constraints into account, the Accessible Resource is estimated to be about 380 TWh/year.

D. Environmental Aspects

The major environmental benefits of wind energy are that it generates electricity without the consumption of fossil fuels from finite reserves, and without generating by-products such as ash, nitrogen oxides, sulphur dioxide and carbon dioxide.

At present there is still considerable debate about the acceptability of wind farms in the landscape of England and Wales, despite the fact that 50% of the planning applications in the first two rounds of the NFFO were approved. The response of the planning authorities has been varied, mainly because of lack of precedent to guide decisions. To help in the assessment of projects, the Department of the Environment, the DTI and the Welsh Office have published Planning Policy Guidance for renewable energy developments. Development issues relating to the Scottish resource may be rather different because of the population distribution. In particular, increased consideration may need to be given to alternative land usage and high scenic value.

Environmental issues will restrict the exploitation of the onshore wind resource but offshore wind energy appears to have few environmental problems provided that areas needed, or already used, for other purposes are avoided.

The environmental burdens resulting from wind energy generation are summarised in table 2.

The savings in emissions arising from the use of wind-powered plant are shown in table 3. It is assumed that each kWh generated by wind power displaces a kWh generated by the average plant mix in the UK generation system (for the year 1990) and thereby saves its associated emissions.

Environmental burden	Specification
Gaseous emissions	None
Ionising radiation	None
Wastes	No fuel wastes
Thermal emissions	None
Amenity/Comfort:	
Noise	Compliance with Local Authority requirements
Visual intrusion	Significant impact

Table 2: Environmental burdens associated with wind energy generation.

Emission	Displaced emissions (grammes of oxide per kWh)	Annual saving for 1 MW scheme (tonnes of oxide)
Carbon dioxide	734	2,200
Sulphur dioxide	10	30
Nitrogen oxides	3.4	10

Table 3: Emission savings relative to generation from the average UK plant mix in 1990.

The major environmental impacts of land-based wind turbines are discussed below.

Land Utilisation

Wind turbines occupy very little ground space, so they can co-exist with livestock or even cereal crops. The only permanent usage of land is for the concrete foundation and the service road.

Visual Impact

For good wind speeds, wind farms need to be sited on high exposed land and so the turbines, together with the power transmission lines from the site, can be visually intrusive. The assessment of visual impact of wind turbines is somewhat subjective. There is a delicate balance to be struck between some loss of landscape value and the environmental benefits of the technology. A wind turbine is not a particularly large structure, typically having a tower height of 25 to 30 m and a rotor diameter of 30 to 35 m. However, wind farms proposed for the UK consist of 10 to 30 machines distributed over several square kilometres of high, open land giving rise to possible conflicts over landscape values.

Noise

Noise from wind turbines is less than from many other everyday activities. Two types of noise are generated by a wind turbine - aerodynamic, from the blades, and mechanical, from the rotating machinery. Aerodynamic noise has been likened to the swishing sound caused by branches of trees during a brisk wind. Mechanical noise can be minimised using well proven engineering practices. Because of the nature of sites required for wind farms, wind turbines will often be located in areas of low background noise where the added noise contribution from the turbines may be detectable. Even in these circumstances, it is likely that the wind turbine noise will only be detectable for limited periods during low wind speed operation; at higher wind speeds the ambient noise level due to wind noise from trees and buildings may increase sufficiently to mask the turbine noise. Careful design, siting and operation should ensure that wind turbine noise is not a nuisance. Noise considerations, however, may limit the number of turbines which can be sited in a specific area. Figure 4 shows some typical noise levels from everyday sources.

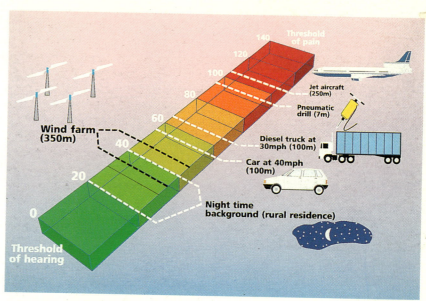

Figure 4: Chart of some common noise levels.

Electromagnetic Interference (EMI)

Interference with television reception is not usually a problem and any necessary remedial action is simple and cheap. Other impacts such as interference with public service transmissions (e.g. microwave links) are very site-specific. Sites likely to cause problems of this kind would be avoided.

Power Transmission

The choice of power transmission route through scenic areas may also prove to be sensitive but there are well established procedures for handling this issue, based on many years of experience from power line planning applications.

Ecology

There are some concerns about the effects of wind turbines on local ecology. These usually focus on land usage, fragile ecosystems (such as peat land and flora) and the potential effects on birds. In fact the amount of land used in constructing a wind farm is only a few square metres per machine and the effect on natural habitat and farming activities is minimal. Fragile land might be damaged during construction of the wind farm but adverse effects can usually be mitigated through good construction practice. Birds could be affected either through displacement from existing habitat or through impact with rotating blades but experience shows that the effects of wind turbines on bird life are small.

Health

The only risk to the public arises from low probability events in which structural failure of a machine would cause pieces to break off causing an airborne hazard.

E. Economics

The cost of energy from wind turbines is site-specific and therefore the costs given here are only indicative and subject to considerable variation. The costs are those applying in 1992 and have been determined from a variety of sources including a survey of commercially available, series production wind turbines rated at 400 kW typical of those currently used in commercial wind farms. One such machine is shown in figure 5.

Capital Costs

Manufacturing (ex-factory) costs can vary by as much as 20% depending on how well the manufacturer is established in the market. The installed costs of wind energy plant are very site-specific, usually related to the terrain and distances from grid lines and access roads. It is thus not possible to give a single estimate of the likely cost of a scheme. Capital costs are often quoted 'per installed kW' and therefore depend on the method used to rate the machine. A better figure based on the blade swept area allows costs to be compared with the annual energy production. The quoted capital cost is for commercial machines in series production. An indicative capital cost for turnkey contracts to supply, install and commission a wind farm is £425 ± 100 per square metre of rotor swept area or £800 to £1200 per installed kilowatt. Additional costs are incurred in seeking planning permission, setting up the financial package and managing the scheme.

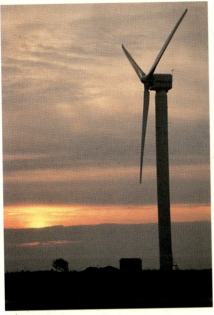

Figure 5: Typical 400 kW wind turbine at Delabole, Cornwall (*photograph courtesy of P. Edwards.*)

Annual Costs

Annual operating costs are much more easily quantified. Operation and maintenance costs average about 0.5 p/kWh. It is assumed that reactive power demand is equal to one half of the energy produced by a wind turbine and reactive power is currently charged at around 0.3 p/unit.

The cost centres associated with establishing a wind farm are given in table 4 along with best values in the UK.

Plant Type: 400 kW wind turbine		
Project initiation	Financing	1% Project cost
	Planning consent	£5,000 to £25,000
	Project development/management	£25,000
Capital costs	Ex-factory cost of the machines	£250 (±25)/sq.m swept area
	Install and commissioning costs	15 (± 5)% ex-factory cost
	Site infrastructure & connection costs	45 (±15)% ex-factory cost
Annual costs	Operation and maintenance costs	1.5 (±0.5)% of capital cost
	Metering and reactive power charges+	0.29 p/kVArh
	Land rental	1.5 (±0.5)% of gross revenue
	Rates	£6.045 per installed kW
	Insurance	0.5% of capital cost

Table 4: Data for calculation of costs for wind farms.

Figure 6: Manufacturing a wind turbine blade
(photograph courtesy of Wind Energy Group Ltd).

Future Costs

It is believed that costs will decrease and performance will increase with time leading to the cost of energy from wind power decreasing to 75% of its 1992 cost by the year 2005, to 70% of its 1992 cost by the year 2025, then remaining constant thereafter. The improvements to the year 2005 are based on historic trends; beyond 2005 the improvements are based on more speculative assumptions such as continuing market growth, accumulated experience and improvements through innovation. The manufacture of a modern wind turbine blade is shown in figure 6.

Energy Output

The energy available at a site depends strongly on the annual mean wind speed. Figure 7 shows an estimate of the annual energy capture (measured at the terminals of the generator) for an isolated 400 kW machine as a function of annual mean wind speed at hub height, assuming 95% availability of the machine. This dependence of energy production on wind speed means that the unit cost of electricity generated by wind plant varies widely with location.

Typically a 400 kW machine operating in a wind farm on a site with an AMWS of 7.5 m/s at 25 m AGL should have a net energy output of around 1,000 MWh/year. This is about 10% less than that produced by an isolated machine, due to clustering effects and electricity transmission losses. The technical factors which determine the annual energy production of the turbine (and hence the unit cost of electricity by combination with the costs) are summarised in table 5.

Performance data:		
Energy capture	Site mean wind speed (25 m agl)	7.5 (±1.0) m/s
	Seasonal wind speed variation factors	1.15 Dec to Feb
	(ratio quarterly/annual mean wind speeds)	1.00 Mar to May
		0.85 Jun to Aug
		1.00 Sep to Nov
Losses	Load factor	30%(±5%)
	Machine availability factor	93%(±3%)
	Array and other losses factors	90%(±5%)
Other data	Construction time	1 year
	Lifetime	20 (±5) years

Table 5: Data for calculation of energy capture by wind farms.

Estimates of seasonal/time load factors are given in table 6 and show that the output from a wind farm should follow seasonal energy demand.

Months	Load factor (day and night)
Dec - Feb	40%
Mar - May	31%
Jun - Aug	22%
Sep - Nov	31%

Table 6: Seasonal/time of day and night load factors.

Cost of energy

Based on the numbers given in figure 7 and table 4, the variation in the unit cost of energy with wind speed has been calculated for a commercially available machine in series production, installed in a medium-sized wind farm of around 20 machines with reasonable access and grid connection. The result is shown in figure 8. The calculation has been carried out for two sets of discount rates (8% and 15% real rate of return) with a 25 year debt repayment, and assuming the capital to be wholly debt financed. These interest rates are chosen as indicative of the range of rates the market might use.

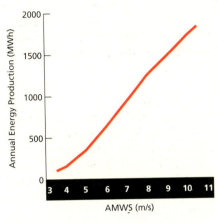

Figure 7: Typical annual energy capture as a function of annual mean wind speed for an isolated 400 kW commercial machine. Delivered output will be less due to losses within the turbine.

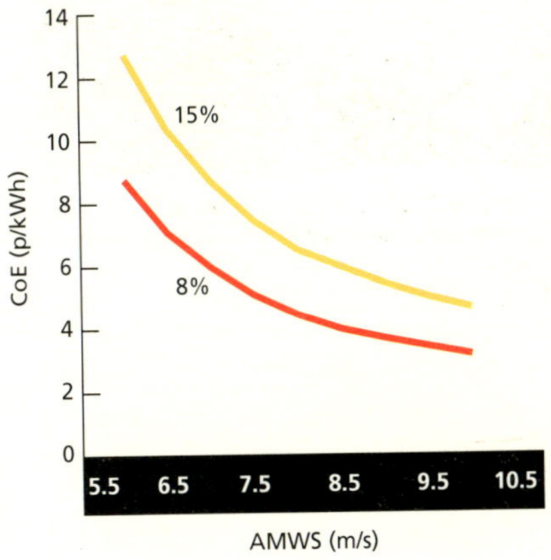

Figure 8: Variation in the cost of energy with annual mean wind speed for 8% and 15% rates of return.

F. Resource-Cost Curves

Onshore

Accessible Resource

The estimates of the Accessible Resource as a function of AMWS are described in section C. It should be noted that data relating to a 33 m diameter, 300 kW machine was used to make these estimates.

Maximum Practicable Resource

The proportion of the Accessible Resource which will be exploited depends on the following.

- **Planning consent.** Not all possible sites are likely to receive planning consent. On the premise that the success rate will be related to the population density, it has been assumed that 10%, 15% and 20% of possible sites would receive planning consent in England & Wales, Northern Ireland and Scotland respectively. On this assumption, the Maximum Practicable Resource is calculated at around 54 TWh/year.

- **Grid Penetration.** Technically there is no specific limit to the amount of non-firm power which can be accepted by the electricity supply system. Before privatisation of the Electricity Supply Industry, the CEGB estimated that up to 20% of the peak power demand on the grid could be provided from non-firm sources without changes being necessary to the grid operating system. More wind power could be accommodated through technical and operational changes, but at increased cost. On the assumption that wind power capacity could provide 20% of the total peak UK demand, the Maximum Practicable Resource in the UK is limited to around 32 TWh/year. This figure assumes that the UK grid is treated as a single system. In practice the level of demand relative to the size of the resource in specific areas of the UK means power will have to be transmitted between areas (particularly Scotland to England) to achieve this usage of power. Details are shown in table 7. At present the transmission of power between the specific areas is limited by the electrical interconnection between them and system strengthening will be necessary if the resource is to be fully exploited.

	England and Wales	Northern Ireland	Scotland
Peak power demand	50 GW	1.5 GW	7 GW
Non-firm power limit	10 GW	0.3 GW	1.4 GW
Energy equivalent	28 TWh/year	1 TWh/year	4 TWh/year

Table 7: Peak demands and assumed penetration limits for non-firm generation.

Total Installed Costs

The total installed cost of a typical UK wind farm of some 20 machines each rated at 400 kW has been calculated from the numbers given in table 4. It is assumed that the costs of installation and commissioning, civil engineering and electrical engineering would be greater for sites with AMWS in excess of 7.5 m/s because of their greater altitude and relative inaccessibility. To take this into account, installed costs have been multiplied by a factor which is 1.0 for sites with an AMWS of 7.5 m/s or less and increases linearly to 1.25 for sites with an AMWS of 10.0 m/s or greater. The total installed cost per machine on a site with reasonable access and grid connection is £400,000 in 1992 prices. The breakdown of this cost is given in table 8.

Capital costs		Annual costs	
Ex-factory cost	£240,000	Operation & maintenance	£6,000
Commissioning & installation	£35,000	Local rates	£2,562
Civil engineering	£35,000	Land rental	£1,350
Electrical engineering	£65,000	Insurance	£2,000
Miscellaneous	£25,000	Reactive power charges	£725
Total installed cost	£400,000	Total annual cost	£12,637

Table 8: Assumed average costs per machine incurred in calculating resource-cost curves.

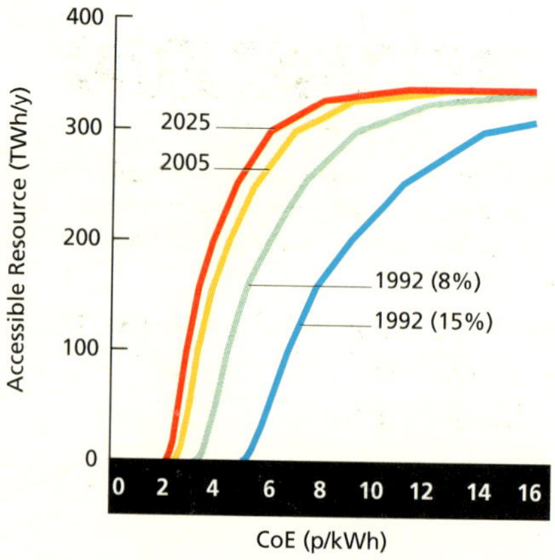

Energy Capture

The energy captured by a wind farm using standard 400 kW production machines (the same as used to produce table 4) has been calculated using the manufacturer's curve of energy production as a function of site AMWS. The curve takes account of machine availability and other performance effects. On the assumption that the turbines are installed in wind farms of about 20 machines, separated by between 7 and 10 rotor diameters, two factors have been applied to reduce the energy capture predicted for a single machine. A factor of 0.95 has been applied to account for electrical losses within the wind farm (e.g. cable and transformer losses) and a further factor of 0.95 has been applied to account for the effect of wake interactions.

Resource-Cost Curves

Resource-cost curves are generated using the costs and the energy capture per wind speed band described above to determine a unit cost for electricity at each wind speed band. The amount of energy which potentially could be captured for each wind speed band is then plotted against the unit cost of electricity (CoE) for that band.

Accessible Resource

Figure 9 shows the estimates of the Accessible Resource in the UK as a function of the CoE for real rates of return of 8% and 15% on invested capital, and how the cost of energy is expected to decrease with time for a real rate of return of 8%. The individual Accessible Resources for England

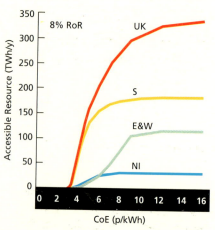

Figure 9: The Accessible Onshore Resource in the UK as a function of cost at 8% and 15% real rates of return (1992 money) and the perceived variation in cost with time at an 8% real rate of return.

Figure 10: The Accessible Onshore Resource for the individual countries in the UK, 1992, at an 8% rate of return (1992 money).

& Wales, Scotland and Northern Ireland are identified in figure 10, also for a real rate of return of 8%.

Maximum Practicable Resource

The Maximum Practicable Resource-cost curve takes into account planning limitations, grid connection costs and the rate at which the wind turbines could be built and deployed. Figures 11 and 12 show the Maximum Practicable Resource for the years 2005 and 2025 at 8% and 15% real rates of return. Expected improvements in wind energy technology lead to a reduction in the cost of energy by 2005 and a further reduction by 2025. The energy contribution from wind will be limited to about 32 TWh/year if no increase in system operating costs is accepted to allow the integration of larger amounts of energy into the existing grid system. This limitation is assumed in the modelling analysis presented in section H.

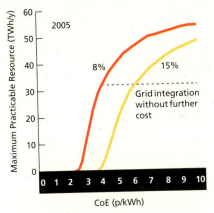

Figure 11: The estimated Maximum Practicable Onshore Resource as a function of the cost of energy for the year 2005 at 8% and 15% real rates of return.

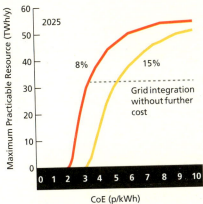

Figure 12: The estimated Maximum Practicable Onshore Resource as a function of the cost of energy for the year 2025 at 8% and 15% real rates of return.

A build rate has been assumed that is sufficient to allow all the resource to be exploited by 2005 so that the total resource available for exploitation in 2005 is the same as that for 2025. This may be optimistic.

Table 9 presents estimates of the resource in the UK available at under 10p/kWh.

	Discount rate (%)	England & Wales	Scotland	Northern Ireland	UK
Maximum Practicable Resource 2005 and 2025 (TWh/year)	8	11	36	5	52
	15	10	35	4	49

Table 9: Estimates of the Maximum Practicable Resource available at less than 10 p/kWh as a function of rate of return.

Offshore

Accessible Resource

The estimate of the Accessible Resource is described in section C.

Maximum Practicable Resource

When the suitability of the sea bed for foundations and alternative sea area usage are taken into account, the resource is reduced to 140 TWh/year. Arguments on the likely penetration of non-firm onshore power supplies into the electricity grid system apply to offshore as well. Offshore and onshore capacities will need to be aggregated when considering overall penetration effects.

Costs

The deployment of currently available commercial machines in offshore wind farms would be considerably more expensive than for corresponding onshore schemes because of the higher foundation and power transmission costs. Paper studies and experience with the Danish and Swedish prototypes indicate that the capital and O&M costs would increase by between 50% and 100% . The cost of energy is likely to be at least 30% higher than current onshore costs.

For the longer term, the Government has commissioned several studies looking at larger machines in very large wind farms. Costs are based on design studies of turbines rated at 6 MW, with a rotor diameter of 100 m, a hub height of 80 m and deployed in large groups of around 330 machines. The estimates of the capital and annual costs for series-produced machines are:

- total installed cost per machine £8,000,000

- annual O&M cost per machine £175,000.

The price base year for these costs is assumed to be 2025 as realisation of such schemes would involve a major development programme on a time scale of 10 to 20 years.

Energy Capture

It is estimated that such a turbine would produce 10.2 GWh/year on a site with an annual mean wind speed of 8.5 m/s at hub height, taking into account machine availability and reductions in individual machine energy capture due to mechanical, electrical and array losses.

ResourceCost Curves

Figures 13 and 14 show estimates of the Accessible and Maximum Practicable Resource for 2025 in offshore UK waters for 8% and 15% real rates of return respectively.

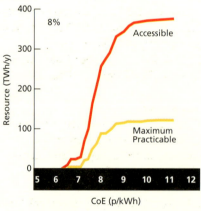

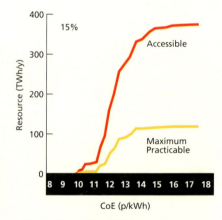

Figure 13: Accessible and Maximum Practicable Offshore Resource in 2025 as a function of the cost of energy for 8% rate of return.

Figure 14: Accessible and Maximum Practicable Offshore Resource in 2025 as a function of the cost of energy for 15% rate of return.

G. Constraints and Opportunities

Technological / System

The UK has one of the best wind resources within the European Union. Onshore wind energy has the potential to make a significant cost-effective contribution to UK energy supplies, possibly as much as 10% of electricity generation. This could contribute to diversity of supply and help reduce carbon dioxide and other emissions from power generation.

The installation of wind energy plant is a growing world-wide business that should allow market opportunities for the UK industry. Over the past 20 years over 20,000 electricity grid-connected wind turbines have been installed around the world. National targets for installed capacity of wind plants have been, or are being, set by several countries - usually several hundreds of megawatts by the end of the century.

The UK has an excellent technology base. Our industry should be capable of exploiting this to produce wind turbines superior to those made abroad. With Government support an infant industry has been created in the UK. This should provide a sound base for the efficient and timely exploitation of market opportunities as they emerge.

Generating plant is likely to be sited in the more remote regions of the country and the transmission of power may require reinforcement of the electrical network, especially between England and Scotland, if the Scottish resource is to be exploited. Most of the lower cost resource is in Scotland and its full exploitation may require the strengthening and extension of the Scottish electricity distribution system. This needs to be taken into account in the future planning of the Scottish Electricity Supply Industry. Further study is required of all system aspects associated with developing the Scottish resource and exporting the power, if rational investment decisions are to be made.

The variability in wind speed over short periods (minutes and hours) is often perceived as a major disadvantage of the resource. In fact, for levels of penetration estimated to be up to 20%, the variation in output from wind plant is less than the statistical variation in demand. Above this level the integration of additional capacity is likely to incur additional system operating costs.

At present, electricity generated by wind power is not economically competitive with that from conventional sources of energy. However, wind energy is still at an early stage in its development. As the technology has matured the cost of wind-generated electricity has fallen. Two avenues for cost reduction can be identified. The first involves changes to manufacturing methods, material preparation and design detail which do not alter the basic nature of the machine. The introduction of series production techniques, rationalisation of supply and stock holding and value engineering fall into this category. The second path to cost reduction involves the application of new knowledge and novel design approaches.

Institutional
Wind farm developments provide a source of local income. This income accrues to the rate payers and the owners of the land used for the wind farm and can have an appreciable beneficial effect on local economies based on subsistence-level farming.

It is usual to employ the local labour force as much as possible in the construction and operation of a wind farm. Such opportunity, whilst seemingly small, can be significant in rural areas with low employment where wind farms are typically sited.

One of the main difficulties encountered in developing wind power in the UK has been the gaining of planning consent from Local Authorities. Visual intrusion has been the primary concern. The best wind sites occur on high, exposed ground and are necessarily remote from conurbations. Such areas are often environmentally sensitive and applicants must satisfy the public, and a variety of special interest groups, that the development will have minimal impact on the environment. Common experience is that as the number of wind installations rises in a country, the difficulty of obtaining consent for further installations also increases. Formal studies to examine the land usage and limitations imposed on the potential resource have been undertaken by several countries including the UK.

The presence of wind turbines will increase public awareness of renewable energy and of the Government's efforts to promote its use. This will help to ease the way for future developments and increase public understanding of energy issues. Several public inquiries have been held on planning applications; these have demonstrated the need for more understanding of the likely environmental impact of wind turbines.

Wind energy is seen as novel and most developers are faced with raising substantial funds against expectations of high rates of return for high perceived risk. However the technology is being exploited world wide, particularly in USA, Denmark and Germany where commercial markets have already been established. As experience is gained it is anticipated that the cost of project finance will decrease.

It can be inferred from published data that wind energy costs will continue to fall in the future. With time, there should be some decrease in manufacturing costs and some increase in the technical performance of plant. These improvements, along with the fact that as experience is gained the costs of installing, commissioning, operating and maintaining wind farms will decrease, make it reasonable to assume that the overall cost of wind-generated electricity will fall by at least 25% by the year 2005.

H. Prospects Modelling Results

Scenario	Discount Rate	Contribution TWh/year			
		2000	2005	2010	2025
HOP	8%	0.16	0.16	0.16	0
	15%	0.16	0.16	0.16	0
	Survey (10%)	0.16	0.16	0.16	0
CSS	8%	0.16	0.16	0.16	9.2
	15%	0.16	0.16	0.16	0
	Survey (10%)	0.16	0.16	0.16	0
LOP	8%	0.16	0.16	0.16	0
	15%	0.16	0.16	0.16	0
	Survey (10%)	0.16	0.16	0.16	0
HECa	8%	28	30	30	30
	15%	9.2	15	15	18
	Survey (10%)	21	24	24	24
HECb	8%	28	30	30	30
	15%	9.2	12	12	18
	Survey (10%)	21	24	24	24

Table 10: Potential contribution from onshore wind under all scenarios.

Data based on the Maximum Practicable Resource-cost curves presented in section F have been used as input for a series of energy model calculations covering a range of scenarios and discount rates to determine the contribution wind energy might make in the year 2025. These are summarised in tables 10 to 12.

Environmental burden			Incremental contribution TWh/year			
	Constraint	Base case scenario	2000	2005	2010	2025
Carbon dioxide	- 5%	CSS	0	0	0	16
	- 5%	HOP	0	0	0	15
Carbon dioxide	- 10%	CSS	0	0	0	24
	- 10%	HOP	0	0	3.6	24
Sulphur dioxide	- 40%	CSS	0	0	0	3.7
	- 40%	HOP	0	0	0	0
Nitrogen oxides	- 40%	CSS	0	0	15	24
	- 40%	HOP	0	0	0	3.7

Table 11: Incremental contribution from onshore wind under environmental constraints. (TWh/year above base case scenario contribution.)

Incremental contribution (TWh/y)				
Discount rate/year	2000	2005	2010	2025
8%	3.6	3.6	3.6	8.6
15%	0	0	0	9.2
Survey	0	0	0	21

Table 12: Incremental contribution from onshore wind under Shifting Sands assumptions. (TWh/year above CSS contribution.)

Examination of the tables shows that onshore wind power is expected to make a significant contribution only for scenarios which reflect Heightened Environmental Concern. For all other scenarios the only contribution will be from capacity installed under the NFFO.

For the HEC scenarios the uptake of capacity approaches the level of 32 TWh/year above which it is assumed additional operating costs would be incurred. It could be argued that even if these extra costs were included in the resource-cost curves, there would be further uptake of onshore wind power.

Due to its higher costs, offshore wind energy does not contribute under any scenario.

II. Programme Review

A. United Kingdom

Wind energy is a well established technology, and the introduction of the NFFO has allowed demonstration on a large scale with over 115 MW of wind power currently installed in England and Wales. The major barrier to true commercial exploitation is the cost of the generated power which is higher than that from conventional sources. However, this is largely due to the relative immaturity of the technology. The payment of premium prices for energy produced is giving industry the opportunity to reduce costs through experience and take advantage of series production, value engineering and other technical improvements.

The Government's Programme
In the U K, the potential of wind energy for the large scale generation of electricity has been under systematic investigation for over 15 years through

programmes involving the Department of Energy (now the Department of Trade and Industry), the Electricity Supply Industry and the manufacturers. The aim of the programmes has been to take the development of wind energy technology for large scale electricity production through the demonstration phase and on towards commercialisation. To date the programme strategy of the DTI (DEn) has involved:

- construction, operation and monitoring of prototype wind turbines to explore the technology;

- extensive underlying R&D programme;

- investigation of the potential for the development of larger machines;

- support of selected wind farms under the NFFO to obtain data on the environmental, technical and economic aspects of commercially available machines and to obtain system operating experience.

In addition the DTI has an ongoing programme to assess the available resource and the likely cost of exploiting it.

B. Other Countries

World-wide, interest in and commitment to the development of wind power generation continue to increase. Many nations involved in the development of wind energy have announced increased targets for installation (usually several hundreds of megawatts by the end of the century) and increased funding for the exploitation of the technology. Leading countries are Denmark, 2% of whose electricity needs are generated by wind turbines, the USA, where over 2,000 MW are installed and Germany, where 230 MW are currently being installed. Nowhere is the technology cost-competitive with conventional power technology, but development programmes by most of the industrialised countries are steadily reducing costs.

Hydro Power

I. Technology Review

A. Status of the Technology

Hydro power has been used as a source of energy for centuries – primitive devices from as far back as the first century BC have been found. Hydro power was a primary source of mechanical energy in the UK before and during the Industrial Revolution but its use has declined since then because of advances in other energy technologies notably the centralised generation of electricity by coal combustion. One of the first large-scale hydroelectric schemes was built in Scotland in 1896.

Figure 1: Belper Mill, Derbyshire: site of a refurbished small-scale hydro scheme.

Hydroelectric schemes can for convenience be divided into two broad categories: large-scale and small-scale. Large-scale schemes will be taken here as having an installed generating capacity of greater than 5 MW. The assumption is that large-scale schemes would be developed and operated by major electricity utilities and are often hundreds of megawatts in capacity, usually involving a dam. Small-scale schemes are usually operated by private developers, estate owners, small companies and some Regional Electricity Companies. Examples are historic water mills consisting of a small weir and a diversion by a pipe or leat, or a buried pipe from a mountain lake to a turbine house. Systems of a few tens of kilowatts are often referred to as 'micro hydro' plants; these are not usually connected to the grid.

The principles of operation of small-scale and large-scale hydroelectric schemes are essentially the same. A scheme consists of the following items:

- a suitable rainfall catchment area;

- a hydraulic head;

- a water intake placed above a weir or behind a dam;

- a method of transporting the water from the head to a turbine, such as a penstock or leat;

- a flow control system;

- a turbine, a generator, associated buildings and grid connection;

- an outflow, where the exhaust water returns to the main flow.

Large-scale schemes typically also include a water reservoir providing daily or seasonal storage to match the production of electricity to the demand for power from the National Grid.

Some hydroelectric schemes can be used as an energy storage device. Such pumped storage schemes usually require two similar-sized reservoirs at different heights. When available, excess energy from the grid is used to pump water from the lower to the higher reservoir. In times of high electricity demand, water flows from the upper reservoir and is converted to electricity as in a conventional scheme. If there is a net water flow into the upper reservoir, a pumped storage scheme may also have a net generating capacity. Examples of pumped storage schemes exist at both small and large-scales. In isolated communities with a local grid, small hydro pumped storage schemes have been used to store surplus energy from wind turbines.

There is a large variation in the combinations of flow and head at sites used for hydro power. Because of this the generating capacities of existing installations vary from less than 1 kW to more than 1,000 MW. The status of hydro power technology is summarised in table 1. The location of existing schemes is very varied – from converted textile mills in lowland town centres to remote Scottish streams.

Current status	Small-scale commercial	Large-scale commercial
UK installed capacity	~80 MW	1,360 MW
World installed capacity *All hydro power*	~550 GW or 2,040 TWh/year in 1988	
UK industry	Several well established UK equipment suppliers	

Table 1: Status of technology.

Installations with a high hydraulic head typically have a pipe (penstock) to convey water to the turbine. If this is buried, the most visible part of a scheme will be the turbine and generator house which can be designed in an acceptable style using local building materials. For a 100 kW scheme, the floor area of a turbine house with a transformer needs to be around 20 square metres. Installations with low hydraulic heads tend to be more visible, as the turbines are of larger dimensions and volumes of water flowing are much greater. Such an installation would typically consist of a weir supplying an open leat or channel running to a turbine house. Hydroelectric schemes installed within the existing water supply industry infrastructure cause negligible additional visual or environmental intrusion as they often use existing buildings. A scheme that involves the impoundment of a large volume of water in a reservoir has a more significant impact on the locality.

Hydroelectric schemes have long operating lives. The civil engineering works (i.e. weirs and water channels) for large and small schemes can last for decades with suitable maintenance. The mechanical and electrical plant has a life from 15 to 50 years. The decommissioning of a redundant hydro power scheme does not pose any problems.

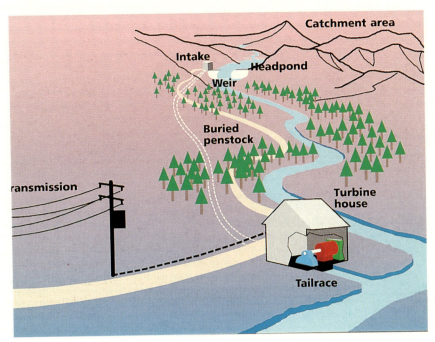

Figure 2: A typical small-scale hydro scheme.

Turbines can be described as either impulse or reaction types. Generally, impulse turbines are high head, low flow rate machines such as the Pelton and Turgo wheels. Reaction machines operate at lower heads and higher flowrates, examples being the Kaplan and bulb turbines. The middle ground is held by the Francis and cross flow turbines, although overlap exists depending upon engineering and economic considerations.

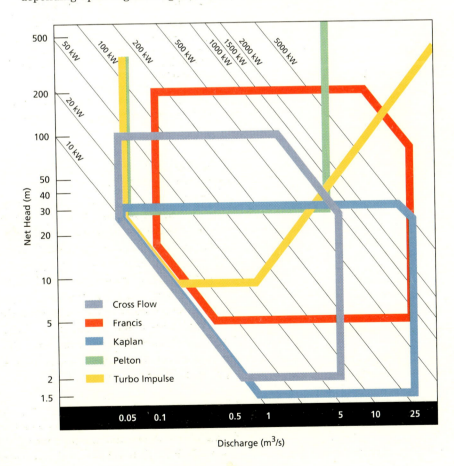

Figure 3: Diagram of net head vs. flow rate for hydro turbines.

Figure 3 shows a typical turbine selection chart as a function of water flow and head. For very small-scale applications, commercial pumps may be run in reverse as low cost turbines. In the UK there are several turbine manufacturers, notably Gilbert Gilkes & Gordon, Biwater Hydropower and New Mills Hydro. Overseas there are many manufacturers including companies such as Ossberger, Bouvier and Fuji.

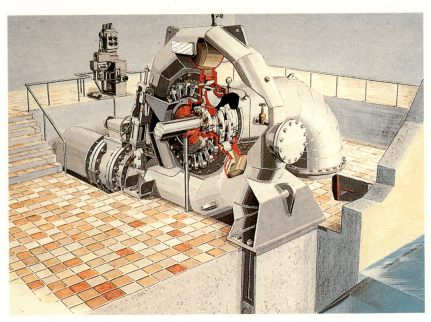

Figure 4: Cutaway of Francis Turbine (Courtesy Gilbert Gilkes and Gordon).

B. Market Status

In the UK, hydroelectric power accounts for around 2% of the total installed generating capacity. Globally, hydroelectric power accounts for around 6% of the world's energy requirements or about 15% of the world's generated electricity. The proportion of a country's energy derived from hydroelectric power is dependent upon the topography, climate, stage of development and availability of other local fuels; this leads to wide variations throughout the world. The need for cheap electricity to refine metals such as aluminium and copper has influenced the development of large-scale hydroelectric schemes in some countries.

Hydro power technology is fully commercialised. Turbines, plant, engineering services and turnkey systems are sold by UK and overseas organisations.

Most of the UK's large-scale hydro capacity is installed in Scotland. Most of these Scottish schemes are owned and operated by Hydro-Electric plc (the privatised successor to the North of Scotland Hydro-Electric Board NSHEB) or Scottish Power plc the former South of Scotland Electricity Board. In England and Wales most hydro installations (both small and large-scale) belonged to the former Central Electricity Generating Board. These sites are now owned by the successor companies – National Power, Nuclear Electric, PowerGen and National Grid Company.

The small-scale schemes in Scotland belong mostly to Hydro-Electric plc, Local Authorities (as part of their water supply system), estate owners and private individuals.

The data presented in table 3 includes the former nationalised Electricity Board sites and schemes contracted under the first two tranches of the NFFO for water companies, Regional Electricity Companies and private generators.

Large-scale sites	MW
England	6
Scotland	1,220
Wales	134
Northern Ireland	None
Total large-scale capacity	1,360

Table 2: Large-scale hydro power currently installed in the UK.

Small-scale sites	MW
England & Wales	20
Scotland	58
Northern Ireland	None
Total small-scale capacity	78

Table 3: Small-scale hydro power currently installed in the UK.

Large-scale hydroelectric schemes have not been included in NFFO rounds to date as no new schemes have been constructed since the NFFO was instigated. However, the first NFFO tranche contained 26 small-scale schemes, with a total installed capacity of 11.8 MW. Of these, 12 were new or refurbished developments, with a total contracted capacity of 5.2 MW. The second tranche contained 12 schemes with a contracted capacity of 10.9 MW.

C. Resource

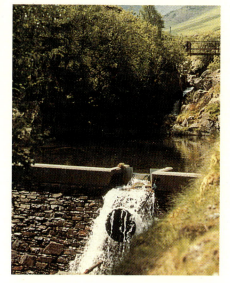

Figure 5: Small-scale hydro scheme at Glenridding in the Lake District.

There is potential for the development of both large and small-scale hydroelectric power in the UK. The major factors that determine the size of the resource are:

- the number of suitable catchment areas;

- the rainfall in suitable catchment areas;

- environmental and social considerations;

- economic considerations.

The total hydro power resource for the UK is estimated at 40 TWh/year or 13 GW of installed capacity (this is based upon mean annual rainfall figures, land area and elevation data). However, a recent study of potential sites in the UK which took account of geographical constraints on potential sites indicates that the Accessible Resource is considerably less than this.

In Scotland, in addition to the existing large-scale capacity of 1.22 GW, there may be an unexploited Accessible Resource of up to 1 GW or 3 TWh/year. This resource would require reservoir storage and its development is therefore likely to be limited by environmental constraints. The unexploited Accessible Resource in Wales has not been quantified but is thought to be small because of environmental considerations. There are no large-scale resources in England or Northern Ireland.

The unexploited Accessible Resource for small-scale hydro in the UK might be in the region of 400 to 700 MW. An additional estimated Accessible Resource of around 500 MW exists at sites with hydraulic heads of less than 3 m, but at present exploitation of this is not likely to be economic. Small-scale schemes particularly those involving the renovation of existing sites are usually less socially and environmentally intrusive than large-scale ones. In general, arranging for a scheme to be acceptable in environmental and social terms will add to its cost, but will not necessarily prevent its development. Table 4 shows the Accessible Resource for large and small-scale hydro power in the UK for two discount rates.

Accessible Resource Measure (TWh/year)	Discount Rate	England &Wales	Scotland	Northern Ireland	UK
Small-scale	8%	0.45	3.4	0.05	3.9
	15%	0.45	3.1	0.05	3.6
Large-scale	8%	0.3	6.6	0	6.9
	15%	0.3	4.5	0	4.8

Table 4: Estimates of the hydro power Accessible Resource for the UK at less than 10p/kWh.

The main constraints on the exploitation of the Accessible Resource are:

- the problems of water abstraction, fishing rights and multiple ownership of the land at weirs and potential river sites;

- environmental constraints in National Parks and wilderness areas;

- the relatively high initial capital costs for low head schemes and the perceived complexity of developing new schemes, planning, water rights etc.

Figure 6: Large-scale hydro dam, Sloy, Scotland (courtesy Scottish Hydro-Electric plc).

D. Environmental Aspects

Hydroelectric schemes do not produce any atmospheric emissions. The turbines and generators create some noise but this is confined to plant normally inside buildings with very thick walls. The noise level from a hydro plant depends on the operating conditions, in particular the installed capacity, hydraulic head and volumetric flow. Small-scale schemes with suitable insulation in the turbine house do not create noise significantly above the ambient background.

The main environmental impacts of hydroelectric schemes are visual intrusion in the landscape and changes to the local ecology. Schemes that do not involve the collecting of water behind dams or in reservoirs have considerably less impact on the environment.

The main environmental concerns are the following:

- the visual intrusion of the water intake, the dam or weir, and turbine building;

- the impact on the river ecology of the diversion of water flow and the need to maintain sufficient flow through the normal river channel;

- any damage to organisms passing through turbines such as fish 'ladders' (linked pools with low barriers) can be provided for fish around hydro schemes;

- the impact of the construction phase of a scheme when temporary dams may be necessary and there is a risk of disturbing the sediment on the river bed and/or depositing construction materials in the water;

- any change in groundwater levels caused by the dam or weir.

The environmental burdens created by a hydro power scheme are summarised in table 5.

The National Rivers Authority has commented that some of the advantages of a small-scale run-of-river hydro scheme are:

- cross flow turbines can entrain air into the water and improve the oxygenation of the river water to the benefit of fish;

- small-scale hydro stations on rivers are usually manned or inspected daily and can act as a valuable monitoring or observation point for water-borne pollutants from other river users;

- the small-scale hydro station in the course of its normal operation collects and removes a large amount of water-borne debris.

Schemes involving the creation of a new reservoir have a potentially greater impact than others, for example:

- the creation of a new aquatic habitat;

- changes in habitats caused by flooding;

- the visual impact of a new dam and reservoir in the overall landscape.

Hydro power schemes do not present any particular safety hazard. Weirs and dams already exist to control flooding and water supplies. Hydroelectric plant is always guarded by grids or 'trashracks' to prevent damage to the turbines by driftwood and leaves, and consequently present no additional hazard to water users.

Environmental burden	Specification
Gaseous emissions	None
Ionising radiation	None
Wastes	None
Thermal emissions	None
Amenity/Comfort:	
Noise	Negligible and always contained within a building
Visual intrusion	Some small to significant impact as size of scheme increases
Use of wilderness area	Depends on location of water flow and use of dams and on form of grid connection used
Other	None

Table 5: Environmental burdens associated with hydro power.

For the calculations in table 6, it is assumed that each kWh generated by hydro power displaces a kWh generated by the average plant mix in the UK generation system (for the year 1990) and thereby saves its associated emissions.

Emission	Displaced emissions (grammes of oxide/kWh electrical)	Annual savings per typical 1 MWe scheme generating 5 GWh/year (tonnes of oxide)
Carbon dioxide	734	3,670
Sulphur dioxide	10	50
Nitrogen oxides	3.4	17

Table 6: Emission savings relative to generation from the average UK plant mix in 1990.

E. Economics

The key factors affecting the economics of a hydro power development are:

- a large initial capital outlay;

- a long lifetime for the scheme;

- high reliability and availability;

- low running costs;

- no annual fuel costs.

The cost of generation by existing large-scale schemes is estimated to be between 1 and 2.5 p/kWh, calculated on an historic costing basis. The economics of small-scale schemes are easier to derive, and have been the subject of recent study. In general the initial costs are lowest for sites with high hydraulic heads and increase as the hydraulic head decreases. Scheme viability decreases as the generating capacity decreases. Table 7 shows a range of illustrative costs for available plant.

		Head 5 to 225 m Power 5 – 1300 kW	Example plant at high head of 223 m 200 kW
Capital cost		£600 to £3,000/kW	£950/kW
	Plant ex-factory costs	£530/kW	£45,000
	Commissioning and installation	£170/kW	£52,000
	Civil engineering	£540/kW	£40,000
	Electrical engineering	£175/kW	£42,000
	Miscellaneous	£100/kW	£11,000
Annual recurring costs:	Operation & maintenance	£12 to £60/kW	£6,000
	Local rates	£8 to £13/kW	£8 to £13/kW
Performance data:	Plant availability	95%	95%
	Average load factor small-scale	15 to 95%	48%
	Type of operation	Predictable on seasonal cycle	Predictable on seasonal cycle
Other:	Construction time	1 to 2 years	1 to 2 years
	Lifetime	25 to 40 years	25 to 40 years

Table 7: Cost and performance estimates for small-scale hydro power.

The initial costs are, however, very dependent on the details of a particular scheme. Land prices and rents vary according to local circumstances. If existing engineering structures are used, the capital cost can be considerably reduced. Schemes constructed within the existing water supply infrastructure have different cost centres, but overall their costs are comparable to other schemes. The cost and performance estimates in table 7 are taken from recent feasibility studies. Figures are given for a high hydraulic head, high capacity scheme and a range of other schemes.

F. Resource-Cost Curves

Small-Scale Resource

Figure 7 shows the small-scale hydro power Accessible Resource for the UK for discount rates of 8% and 15%. The lower area on the curve represents the installed small-scale resource.

The data for these curves have been derived from a resource assessment study conducted for the Department of Energy between 1987 and 1989. The assumptions made in the study were:

- a 40 year lifetime for the scheme;

- a one year construction period for the scheme;

- an availability of 95%;

- a load factor of 56%;

- an annual operating and maintenance cost which is 2% of the capital cost.

The study concluded that, at a rate of return on capital of 10% or more, the unexploited resources in the UK which could produce power at the 1989 average electricity purchase tariff of 2.7p/kWh were as follows.

- England   19.3 MW

- Wales   14.8 MW

- Northern Ireland   2.2 MW

- Scotland   286 MW

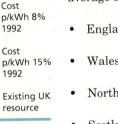

Figure 7: Small-scale hydro power Accessible Resource for the UK. The lower area on the curve represents the installed small-scale resource.

The study covered sites in the capacity range of 25 kW to 5 MW. Both run-of-river and schemes within the water industry were considered as well as conventional hydro. Sites with heads of less than 2 m with existing civil engineering schemes and sites with heads less than 3 m with no civil works were not considered.

The results of the study have been converted to 1992 prices and used in the present assessment to produce the Accessible Resource estimates in figure 7. The lowest costs (<3p/kWh) are likely to be for sites with existing civil engineering works.

The Accessible Resource for small-scale hydro has been derived using data from a site by site assessment of the UK hydro resource which took account of likely constraints on the development of potential sites. This suggests that the whole of the Accessible Resource could be developed if so desired. For the purposes of this assessment, for small-scale hydro the Maximum Practicable Resource and the Accessible Resource are taken to be the same.

Large-scale resource

An attempt has been made to produce a simple resource cost curve for the unexploited large-scale hydro resource. The following assumptions have been made:

- capital costs of £1,000 to £3,000/kW, this is based on inflated historic costs and the mean cost of three large international schemes (200, 650 and 1,600 MW);

- a load factor of 35% – based on the present UK average large-scale load factor;

- an availability of 95%;

- operation and maintenance costs of 0.25% of the capital cost;

- an estimated unexploited resource of approximately 1 GW is divided up, for simplicity, into three capital cost bands (£1,000, £2,000 and £3,000/kW) each of capacity 333 MW;

- a construction period of four years.

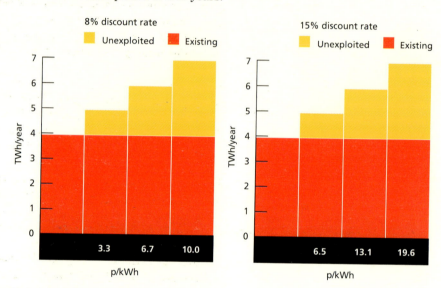

Figure 8: Large-scale hydro cumulative Accessible Resource-cost curve at 8% & 15% discount rates.

Figure 8 presents a resource-cost histogram based on these assumptions for 8% and 15% discount rates. The existing large-scale hydro resource for Scotland and Wales is shown as a continuous component in the bottom part of the diagram.

The unexploited large-scale Accessible Resource may be as large as 1 GW but is likely to be limited because of environmental constraints. The unexploited large-scale Maximum Practicable Resource is considered to be small.

G. Constraints and Opportunities

Large-scale

Technological
The technology for large-scale hydroelectric power is well established. There are in general no insuperable barriers to its deployment that are purely technical in nature. Recent progress in scheme design with water courses being built underground in tunnels has resulted in a reduction of scheme costs. It is probable that the current UK generating capacity could be increased by the refurbishment of existing plant. Because of the unique role of pumped storage schemes, that of transforming base-load generation into peak supply, there is perhaps scope for further deployment of this technology.

Institutional
There are a number of institutional constraints on the further deployment of the large-scale hydro power resource. The large capital cost and long payback period which characterises large hydro schemes make it difficult for new

schemes to compete with existing thermal plant. The environmental and social disruption associated with large-scale hydro power can result in new schemes failing to receive planning permission. At present there is an excess of generating capacity in Scotland where most of the large-scale resource is located. There is also a limit on the transmission capacity from Scotland to England and Wales, so there is little incentive for the creation of new capacity.

Small-scale

Technological
The high cost of turbines compared to the civil engineering costs for a scheme at hydraulic heads of 3 m or less restricts the deployment of the technology at this end of the spectrum. Many potential small hydro sites are in remote areas where there is no local demand. This makes them uneconomic to develop because of the costs of grid connection and transmission. Recent work on novel systems for generation at low hydraulic heads indicates that it is unlikely that any new approaches will provide cheaper power than could be achieved using current commercially available equipment. The reduction in the cost of conventional equipment offers the best route forward.

Institutional
Although charges are no longer made for water abstraction, the necessary planning steps to secure the use of water for hydro power may still hamper developments. Securing the necessary land for the development of a scheme through purchase or rental at a reasonable price poses a serious problem in some cases.

Until the introduction of the NFFO, the lack of a secure market for the sale of electricity has discouraged the development of new schemes for grid connection.

H. Prospects Modelling Results

Tables 8, 9 and 10 illustrate the potential contribution from small and large-scale hydro power under the different scenarios used in the MARKAL analysis. The substantial amount of existing installed capacity, particularly in Scotland, provides a base level of over 4 TWh/year throughout all scenarios. The more favourable scenarios for renewable energy sources, the HEC scenarios, predict a maximum energy production of over 6 TWh/year with a large proportion of the Maximum Practicable Resource being developed (up to 50% of the resource below 10p/kWh).

Scenario	Discount rate	Contribution, TWh/year			
		2000	2005	2010	2025
HOP	8%	4.2	4.2	5.1	5.1
	15%	3.9	4.2	4.2	4.2
	Survey (10%)	4.2	4.2	4.2	5.1
CSS	8%	4.2	4.2	4.2	5.1
	15%	4.2	4.2	4.2	4.2
	Survey (10%)	4.2	4.2	4.2	5.1
LOP	8%	4.2	4.2	4.2	4.2
	15%	3.9	4.2	4.2	4.2
	Survey (10%)	4.2	4.2	4.2	4.2
HECa	8%	6.2	6.2	6.2	6.2
	15%	5.1	5.1	5.1	5.1
	Survey (10%)	6.2	6.2	6.2	6.2
HECb	8%	6.2	6.2	6.2	6.2
	15%	5.1	5.1	5.1	5.1
	Survey (10%)	6.2	6.2	6.2	6.2

Table 8: Small and large hydro combined potential contribution under all scenarios.

Environmental burden			Incremental contribution (TWh/year)			
	Constraint	Base case scenario	2000	2005	2010	2025
Carbon dioxide	- 5%	CSS	0	0	0	0
	- 5%	HOP	0	0	0	0
Carbon dioxide	-10%	CSS	0	0.97	0.97	0
	-10%	HOP	0	0.97	0.97	0
Sulphur dioxide	-40%	CSS	0	0	0	0
	-40%	HOP	0	0	0	0
Nitrogen oxides	-40%	CSS	0	0	0.97	0
	-40%	HOP	0	0	0.97	0

Table 9: Small and large hydro combined: incremental contribution under environmental constraints (TWh/year above base case scenario contribution).

Incremental contribution (TWh/year)				
Discount rate	2000	2005	2010	2025
8%	0.97	0.97	0.97	0
15%	0	0	0	0.97
Survey	0.97	0.97	0.97	0

Table 10: Small and large hydro combined: incremental contribution under Shifting Sands assumptions (TWh/year above CSS contribution).

II. Programme Review

A. United Kingdom

Hydroelectric technology is well established in the commercial market. Manufacturers of turbines, control systems and generators have been trading for many years both in the UK and abroad. The principal barrier to further development of hydroelectric schemes for export of electricity to the grid is the absence of a secure market. Conventional large-scale developments are discouraged by institutional and economic factors. The bulk of the large-scale resource is in Scotland where an excess of generating capacity makes new developments unattractive in the current economic climate.

However, government has taken initiatives to encourage the development of small-scale hydro. These include:

- the removal of water abstraction charges;

- the rationalisation of the local authority rating system;

- the provision of discretionary partfunding of feasibility studies;

- the creation of an improved electricity market through NFFO support.

The hydro industry has a role in developing both new and derelict sites as schemes for the NFFO. Many small-scale hydro sites are not exclusive to the hydro industry, the land is often owned by farmers, estate owners, mill owners and the water supply industry. In particular the water industry is developing its spare and unused hydro power potential through the NFFO as part of its water movement and reservoir operations.

The Government's Programme

The Government's programme for small-scale hydro power has been focused on two main areas:

- stimulation of the wider take up of existing commercially available small-scale hydro in relevant market sectors, including water companies, industry and private individual operators;

- support of RD&D on novel technologies for extracting energy from lowhead sources to establish whether or not they are technically or commercially viable.

A technical and economic assessment of the UK resource for small-scale hydro has been completed and the results have been disseminated. These are now being used by developers to locate suitable sites.

A study of the nontechnical barriers to the exploitation and commercial development of small-scale hydro power has been carried out. The results have been published and made available to the relevant parties. Some of the obstacles identified in this study have now been removed, and those that remain have been highlighted. One problem identified by the study is the initial financial outlay necessary to determine whether or not a particular development is feasible. The Government has therefore been contributing 50% towards the cost of feasibility studies for small-scale hydro power sites. A lack of technical and procedural knowledge (planning requirements, electricity sales etc.) of small-scale hydro power on the part of developers was also identified as a reason why schemes failed to be successfully completed. Planning and technical guidance documents are being produced to meet this need.

The programme has also funded a number of research projects on novel concepts for the generation of electricity at sites with low hydraulic heads. This work is now largely complete. None of these innovations has yet proved to be attractive enough to take to the development stage.

B. Other countries

Measures employed by those governments which support the development of hydro power include:

- favourable planning legislation, permitting scheme construction;

- technical assistance for developers;

- low interest (e.g. 5%) loans covering some or all construction costs;

- tax relief on capital cost of very small plant (tens of kW);

- assured sale price for generated electricity;

- guaranteed long term market for generated electricity.

Tidal Power

I. Technology Review

A. Technology Status

Origin of the Tides

Tides are caused by the gravitational attraction of the moon and the sun acting upon the oceans of the rotating earth. The relative motions of these bodies cause the surface of the oceans to be raised and lowered periodically, according to a number of interacting cycles. These include:

- a half day cycle, due to the rotation of the earth within the gravitational field of the moon;

- a 14 day cycle, resulting from the gravitational field of the moon combining with that of the sun to give alternating spring (maximum) and neap (minimum) tides;

- a half year cycle, due to the inclination of the moon's orbit to that of the earth, giving rise to maxima in the spring tides in March and September;

- other cycles, such as those over 19 years and 1,600 years, arising from further complex gravitational interactions.

The range of a spring tide is commonly about twice that of a neap tide, whereas the longer period cycles impose smaller perturbations.

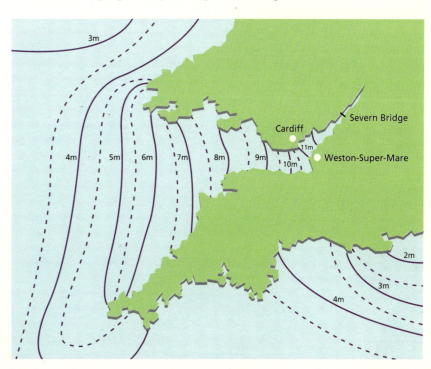

Figure 1: Tidal resonance in the Severn Estuary, UK: mean spring tidal range.

In the open ocean, the maximum amplitude of the tides is about one metre. Tidal amplitudes are increased substantially towards the coast, particularly in estuaries. This is mainly caused by shelving of the sea bed and funnelling of the water by estuaries. In some cases the tidal range can be further amplified by reflection of the tidal wave by the coastline or resonance. This is a special effect which occurs in long, trumpet-shaped estuaries, when the length of the estuary is close to one quarter of the tidal wave length.

Figure 1 shows how these effects combine to give a mean spring tidal range of over 11 m in the Severn Estuary. As a result of these various factors, the tidal range can vary substantially between different points on a coastline.

The Available Energy
The amount of energy obtainable from a tidal energy scheme therefore varies with location and time. Output changes as the tide ebbs and floods each day; it can also vary by a factor of about four over a spring neap cycle. Tidal energy is, however, highly predictable in both amount and timing, in contrast to some other sources of renewable energy.

The available energy is approximately proportional to the square of the tidal range. Extraction of energy from the tides is considered to be practical only at those sites where the energy is concentrated in the form of large tides and the geography provides suitable sites for tidal plant construction. Such sites are not commonplace but a considerable number have been identified in the UK.

The Technology
Tidal Power is one of the older forms of energy used by man. Records indicate that tide mills were being worked on the coasts of France, Spain and Britain before 1100 AD. These consisted simply of a storage pond which was filled during the flood tide through a sluice and emptied during the ebb tide through a waterwheel. They remained in common use for many centuries but were gradually displaced by the cheaper and more convenient fuels and machines made available by the Industrial Revolution.

Tidal Barrage Design and Construction
In the modern version of the tide mill, a barrage constructed across an estuary is equipped with a series of gated sluices and the waterwheel is simply replaced by a bank of low-head axial turbines. Where it is necessary to maintain navigation to the upper part of the estuary, a ship-lock may be required. Figure 2 shows the elements of a typical barrage design.

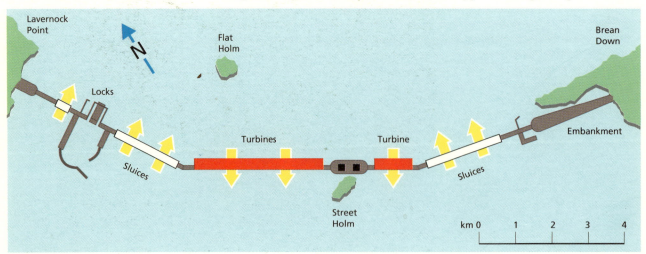

Figure 2: Schematic representation of the proposed Severn Barrage.

The construction method usually proposed for the main barrage structure involves the use of caissons. These are large prefabricated units of concrete or steel, which would be manufactured remotely at suitable construction yards and then be towed to the barrage site and sunk into position on prepared foundations. Each caisson element would house a group of turbine generators or sluices, or simply be blank to make up the remainder of the structure. In shallower water, conventional embankment could be used to complete the structure.

Construction of a barrage 'in the dry' behind a temporary coffer dam, which is later removed, was carried out successfully by the French at La Rance in the 1960s. However, this method is usually more expensive than the caisson approach and, for larger deeper estuaries, is generally considered to be too risky. Also, the complete closure of an estuary by a single impermeable coffer dam would not be considered environmentally acceptable today.

Other methods of building have been considered, including the construction of diaphragm walls of reinforced concrete within a temporary sand island. However, no significant cost advantages over the caisson approach have yet been demonstrated and studies for the Mersey Barrage (UK) have indicated that the use of diaphragm walling could increase construction time by about two years.

Figure 3: Artist's impression of a large tidal barrage turbine.

Electricity is generated using large axial flow turbines, of diameters up to 9 m. In view of the continuously varying head of water which drives them it is necessary to regulate the blade angle of the distributor or of the turbine runner, or both, for maximum efficiency. If the turbine is also to be used in both directions for generation, or in reverse for pumping, variable control of both the distributor and the turbine runner (i.e. double regulation) may be recommended.

Two principal types of turbine generator have been considered. The conventional 'bulb' turbine contains the generator in a pod located in the water passage directly behind the turbine runner. For larger machines, a directly driven multi-pole generator in the bulb is generally preferred (e.g. Severn). A geared bulb turbine in a pit configuration may be cheaper if lower rated machines are required (e.g. Mersey & Conwy).

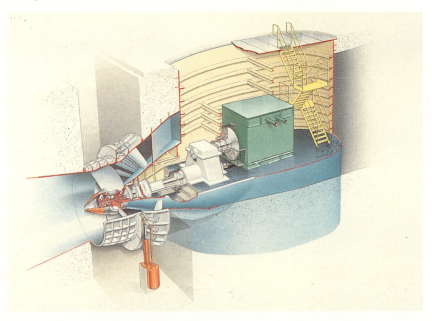

Figure 4: Cutaway of a double-regulated pit turbine with gearbox-driven generator (courtesy Kvaerner Energy a.s.).

A rim-generator turbine, in which the stator is outside the water passageway and the rotor is fixed to the periphery of the turbine runner, has been successfully tested at Annapolis in Canada. However, since large rim-generator turbines are not available with double regulation and therefore no proven capability for reverse turbining or pumping, their use may be restricted to simple ebb generation schemes.

Operation

Ebb generation is the simplest mode of operation for a tidal barrage scheme. The operating cycle consists of four steps:

- sluicing on the flood tide, to fill the basin;

- holding the impounded water until the receding tide creates a suitable head;

- turbining the water from the basin to the sea, on the ebb tide, until the tide turns and rises to reduce the head to the minimum operating point;

- holding until the tide rises sufficiently to repeat the first step.

Ebb generation with flood pumping is a modification of this mode which is presently favoured by UK developers because of the ability to increase energy output. By using the turbines in reverse as pumps, the basin level, and hence the generating head, can be raised. The energy required for pumping must be imported but, since the pumping is carried out against a small head at high tide and the same water is released later though the turbine at a greater head, this can produce a net energy gain with some limited ability to re-time output. UK studies on a number of tidal energy schemes indicate that the energy gain through pumping could be small but useful and typically in the range 3% to 13%. Figure 5 illustrates this mode of operation and the effect on water levels and energy output.

Other modes of operation, including flood generation and two-way generation, have been considered but are not generally favoured for the UK, since they either yield less energy or require more complex plant and machinery. Also their lowered water levels may be less acceptable to shipping.

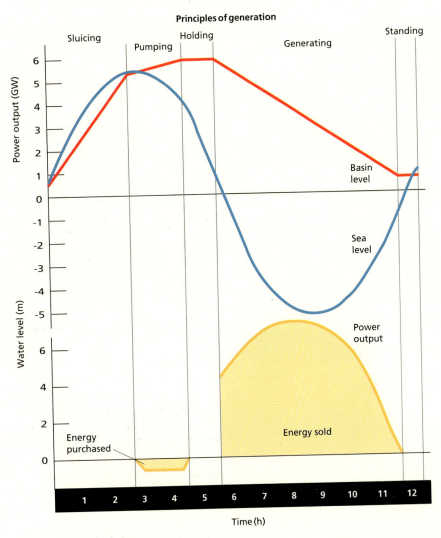

Figure 5: Ebb generation with pumping.

Life expectancy
Tidal energy barrages are expected to have very long lifetimes. Their design life could be about 120 years but, with normal maintenance and replacement of turbine generators at 40 year intervals, their lifetime could be effectively unlimited. Once built, they would not need to be decommissioned.

B. Market Status Existing

Tidal Energy Schemes
Relatively few tidal power plants have been constructed in the modern era. Of these, the first and largest is the 240 MW barrage at La Rance (France), which was built as a full scale demonstration scheme in the 1960s and has now completed 25 years of successful commercial operation.

In 1984 a 17.8 MW plant at Annapolis, Canada, was commissioned to demonstrate a large diameter rim-generator turbine. Others include a 0.4 MW experimental plant at Kislaya Guba, Russia, the 3.2 MW Jiangxia station and a number of small plants in China. There are no tidal energy schemes in the UK.

Operating experience at La Rance and Annapolis has generally been positive. The stators of the La Rance generators and the field windings of the Annapolis generator required initial modification to overcome design weaknesses. However, both plants have since proved very reliable, with availability in recent years of around 97% and with only modest maintenance requirements.

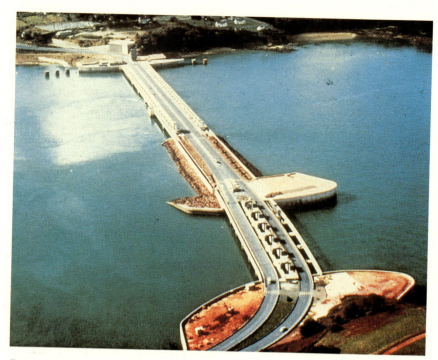

Figure 6: Tidal energy barrage, La Rance, France.

Commercial Status

On the basis of the above schemes, tidal energy can now be regarded as a technically proven and dependable technology. The necessary skills and much suitable equipment are commercially available and several of the largest UK engineering companies are involved in progressing feasibility studies on major projects, such as the Severn and Mersey Barrages.

The main barrier to the commercial uptake of the technology, in the UK as elsewhere, remains the high capital cost of even the best schemes in relation to their energy output, which is consequently reflected in their generating costs. Although output from tidal energy schemes is predictable, the load factor is low at around 23%.

Current status	Overseas: demonstration UK: development
UK installed capacity	0 MW
World installed capacity	France – 240 MW Canada – 17.8 MW China – 5.0 MW
Total	263.2 MW
UK industry	Several well established UK consulting and construction engineering companies and plant suppliers.

Table 1: Status of tidal energy technology.

C. Resource

The UK has some of the most favourable conditions in Europe for generating electricity from the tides, and possesses about half of the total tidal energy resource within the European Union. This is the result of an unusually high tidal range along the west coast of England and Wales, where there is a multiplicity of estuaries and inlets which could potentially be exploited. However, the tidal ranges in Scotland and Northern Ireland are generally too low for economic exploitation.

No estimate is available of the total Accessible Resource from all UK estuaries. However, if every reasonably practicable estuary with a mean spring tidal range exceeding 3.5 m were to be employed for tidal power, the yield would be about 50 TWh/y, representing around 20% of the present electricity consumption in England and Wales. About nine tenths of this would be at eight larger sites (Severn, Dee, Mersey, Morecambe Bay, Solway Firth, Humber, Wash, Thames) yielding 1 to 17 TWh/year each, while one tenth of the potential is at 34 smaller sites, yielding typically 20 to 200 GWh/year each. At 17 TWh/year the Severn Barrage, if built, would be far the largest single renewable energy project and, indeed, one of the largest civil engineering projects in the world.

Table 2 gives estimates of the Accessible Resource for tidal energy which might be generated at a cost of under 10p/kWh, assuming different discount rates. The resource-cost curves shown in figure 8 also show in some detail the sensitivity of the Accessible Resource to the discount rate.

Discount Rate	England & Wales	Scotland	Northern Ireland	UK
8%	19 TWh/year	0	0	19 TWh/year
15%	0	0	0	0

Table 2: Accessible Resource in the UK available at a cost of under 10p/kWh (1992 prices).

The opportunity for harnessing tidal energy in the UK is not restricted to very large schemes, such as the Severn Barrage. Although at 17 TWh/year this represents a large proportion of the most economic potential for tidal energy, tidal power projects of one tenth the size (e.g. the Mersey Barrage, 1.45 TWh/year) or one hundredth the size (e.g. the Wyre Barrage, 0.13 TWh/year) can be identified, which might yield electricity at similar cost.

D. Environmental Aspects

Tidal energy can have a number of environmental and regional effects on the community including a number of amenity benefits which are discussed below.

Savings in Emissions
Once built, a tidal energy scheme is non-polluting, with no emission of noxious gases, ionising radiation, solid or liquid wastes, or heat, (see table 4). Its operation would generally displace the combustion of coal and hydrocarbon fuels by other plant on the network, saving in particular the emission of carbon dioxide, sulphur dioxide and nitrogen oxides. These savings would vary considerably with the size of the scheme. Table 3 shows possible savings in emissions, assuming an average mix of plant would be displaced.

Emission	Displaced Emissions (tonnes oxide per GWh)	Annual savings for typical schemes (tonnes oxide/year)		
		Severn Barrage	Mersey Barrage	Wyre Barrage
Carbon dioxide	734	12,500,000	1,064,000	96,000
Sulphur dioxide	10	170,000	14,500	1,310
Nitrogen oxides	3.4	58,000	4,900	445

Table 3: Emission savings relative to generation from the average UK plant mix in 1990.

For the calculations in table 3, it is assumed that each kWh generated by tidal power displaces a kWh generated by the average plant mix in the UK generation system (for the year 1990) and thereby saves its associated emissions.

Flood Protection

A tidal energy scheme can also provide protection against coastal flooding within the basin during very high tides, by acting as a storm surge barrier.

Amenity Benefits

Other significant non-energy benefits can arise from the building of a barrage, particularly in relation to the regional development opportunities which might be created. These could include the possibility of a road crossing, marina and water sport developments, increased tourism, increased land values, and increased local employment.

Ecology

A tidal energy scheme could cause significant changes to the local estuarine ecosystem. For each scheme a site-specific environmental assessment would be needed to identify these changes and determine their acceptability. In particular, the construction of a barrage would affect the hydrodynamic regime of the estuary, typically reducing the tidal range, currents and inter tidal area within the basin by about half. These hydrodynamic changes can in turn influence both water quality, (for example the dilution and dispersion of effluents), turbidity and the movement and composition of bed sediments. Any reduction in turbidity that might occur could result in increased primary biological productivity, with consequent effects throughout the food chain from phytoplankton to zooplankton, benthic invertebrates, fish, wading birds and wildfowl.

Environmental burden	Specification/extent of burden
Gaseous emissions	None
Ionising radiation	None
Wastes	None
Thermal emissions	None
Amenity/Comfort:	
Noise	Negligible impact
Visual intrusion	Some impact possible
Other	Positive benefits to sailing, water sports, tourism, road transport, employment etc.
Natural environment:	
Hydrology and sedimentation Water quality	Some changes expected, e.g. to basin water levels and currents.
Benthic invertebrates Birds Fish Saltmarsh Nature conservation	The extent and acceptability of ecological changes will need to be determined individually for each site.

Table 4: Environmental burdens associated with tidal energy.

Elements of the estuarine ecosystem, which would need to be characterised when preparing an environmental statement for a site-specific barrage, include the water quality, sediment regime, bird and fish populations and the likely effect on these of the barrage. With careful selection of appropriate sites, suitable barrage designs and operating modes, and limited maintenance dredging, movement of silt need not be a problem. The reduction of inter tidal mud flat areas within the basin (feeding areas for wading birds), may in some cases be compensated by changes in biological productivity leading to increased availability of food. Potential effects on fish, especially migratory species, may arise through mortality at the turbines or through delayed passage, as observed at Annapolis in Canada. However, there have been no significant indications of fish mortality at La Rance, France, or evidence that the population of any fish species has been significantly affected by that plant. Fish passes and diversion systems are available to mitigate such impacts, but further research is necessary to find ways of combining low cost with efficiency.

So far, no extensive monitoring of a tidal energy scheme has been undertaken and more detailed understanding of some estuarine processes is desirable. However, while some environmental uncertainties remain and will require further investigation, especially at local site level for specific schemes, operational experience abroad has been positive. No major factors have been identified which would inhibit adoption of the technology, assuming proper attention to scheme siting and design.

E. Economics

Tidal power incurs high capital costs per kW of installed capacity and construction times can be several years for the larger projects. Operation is also intermittent, with relatively low load factors, typically around 23%. These characteristics tend to increase the unit cost of generation. On the other hand, total annual charges including operation and maintenance costs are very low, typically 0.5% of the initial capital cost. With appropriate maintenance the lifetime of a scheme can be very long, perhaps 120 years for the main barrage structure and 40 years for plant and machinery.

The economics of tidal energy depend primarily on site specific factors, energy output varying with the tidal range and enclosed basin area and costs being related to the barrage length and height plus any special factors such as ship locks to permit continued navigation.

There appears to be no significant economy of scale. UK studies have identified promising schemes with capacities ranging from 30 MW to 8 GW and having broadly similar energy costs, as shown in table 5.

The high capital costs and long construction times of large tidal barrages make their cost of energy particularly sensitive to the discount rate for capital employed, and hence to the method of financing. For example, studies on the Severn Barrage have shown that increasing the discount rate from 8% to 15% would double the cost of energy, from 7.2 to 16.2p/kWh, at 1992 prices.

Tidal energy scheme	Build time in years	Capacity MW	Load factor %	Capacity cost 1992 £/kW	Generation cost 1992 p/kWh 8% DR	Generation cost 1992 p/kWh 15% DR
Severn Barrage	10	8,640	23	1,250	7.2	16.2
Mersey Barrage	5	700	23	1,280	7.1	14.3
Wyre Barrage	2	63.6	24	1,470	6.8	12.4
Conwy Barrage	2	33.4	21	1,610	8.7	15.7

Table 5: Examples of cost & performance estimates from recent barrage studies.
(Notes: costs exclude non-energy benefits; costs for Severn, Wyre & Conwy schemes updated to 1992 prices using RPI; costs for the Severn Barrage do not include an estimated £1B required to reinforce the electricity system to accommodate the output from the project.)

Plant type: Severn Tidal Barrage, 8640 MW				
Investment cost	**£(1992)/kW**	**Annual recurring costs**		**£(1992)/kW/y.**
Capital and installation costs:		Operation and maintenance:		
Civil engineering works	641	Staff		2.71
Contingencies (15%)	96	Materials		1.6
	737	Power for auxiliaries		1.05
		Contingencies		0.66
Turbine generators	346			**6.02**
Contingencies (5%)	17			
	363	Off-barrage costs:		
On-barrage transmission and control	53	Compensation charges and		
Contingencies (10%)	5	environmental monitoring		1.35
	58	Land rental, rates, insurance		3.16
				4.51
Other:				
Feasibility studies	2			
Management and supervision	62			
Parliamentary, planning & environmental approval	7			
Compensation works: land drainage, sea defences, port works, effluent discharge etc.	15			
	86			
Total investment cost	**1,244**	**Total recurring costs**		**10.53**
Performance data:		Other:		
Type of operation	ebb generation	*Construction time*		
Plant availability	95%	To first generation		7 years
Energy availability	97%	To completion of project		10 years
Plant load factor	22.5%	*Expected lifetime*		
Firm power contribution	1,300 MW approx	Barrage structure		120+ years
		Turbine generators		30 to 40 years

Table 6: Cost and performance estimates for the Severn Barrage.
Note: Costs are at 1992 prices (1988 prices x 1.3 change in RPI index) for supply to the barrage boundary and include grid connection, but exclude allowances for grid strengthening, a public road crossing and interest charges during construction.

Season	Day	Night
Spring/Autumn	22%	23%
Summer/Winter	21%	22%

Table 7: Variation in load factor. Notes: plant load factor will vary slightly from scheme to scheme and with mode of operation. The above figures are indicative of a typical barrage operating in ebb generating mode, with flood pumping. Plant load factor will vary from one tide to the next according to the variation in tidal range, which is largely independent of season and time of day. However, on average equinoctial spring tides are slightly greater, and equinoctial neap tides are slightly smaller, than those in summer and winter, yielding on average slightly more energy in the spring and autumn. A barrage may be pumped or the high water stand held, to bring forward or delay generation by up to an hour during the daytime, depending on the time of the tide, in order to maximise the value of energy generated. For the Severn Barrage, it has been estimated that this might lead to an increase in overall value of up to 15%, but a decrease in net energy output, (and hence decrease in daytime load factor) of around 4%.

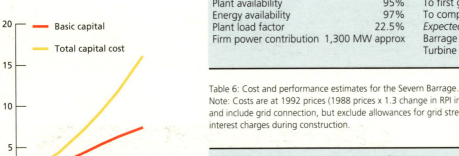

Figure 7: Cost of electricity from the Severn Barrage as a function of discount rate, showing interest during construction.

F. Resource-Cost Curves

Preliminary feasibility studies for tidal energy have now been undertaken on several of the more promising UK estuaries. Capital costs have been obtained using a 'bottom-up' approach. From these the unit cost of generation has been estimated for a range of discount rates.

A parametric formula has also been developed to assess the likely potential of tidal schemes. This relates the unit cost of energy generation to the tidal range and physical dimensions of the scheme, and has been calibrated using data from previous detailed barrage investigations:

$$\log U = k \left\{ \frac{\log L^{0.8}(H+2)^2}{A(R-1)^2} \right\}$$

where:
U = unit cost of electricity (p/kWh)
L = length of barrage (m)
H = maximum height of barrage
       above sea bed (m)
A = area of basin (sq.km)
R = mean tidal range (m)
k = constant

This has been applied to other reasonably exploitable tidal energy sites where the mean spring tidal range exceeds 3.5 m. The potential of over 120 small estuaries and inlets in England and Wales, about 20 in Scotland and two in Northern Ireland has been checked in this way. The results of these two approaches have been combined and are displayed in figure 8.

Using the data for the total UK resource, the potential savings in carbon dioxide emissions have been calculated as a function of energy cost (p/kWh) for a range of discount rates, and are also shown.

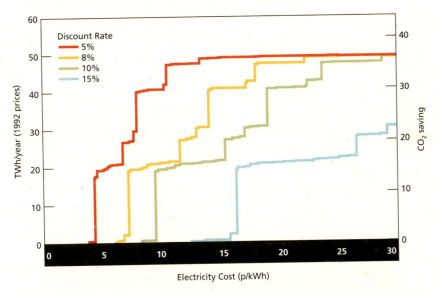

Figure 8: The UK tidal energy Accessible Resource and carbon dioxide savings relative to generation from the average UK plant mix in 1990, for various discount rates.

The resource cost data presented in figure 8 includes sites identified from the parametric assessment as well as data from feasibility studies on the Severn, Mersey, Wyre and Conwy estuaries. Some allowance has been made for the length of time which would be required for detailed design, environmental assessment and parliamentary procedures which would be required before the construction of the Severn Barrage. The whole process would take at least nine years. The resource-cost curve presents the Accessible Resource for tidal power in the UK. Although it would be technically possible to construct barrages simultaneously it is more likely that a single small scale scheme would operate for a number of years before serious investment in other projects, particularly on the scale of the Severn. This would allow sufficient time for effective demonstration of a tidal energy barrage before further deployment at a larger scale. Determination of the Maximum Practicable Resource, based on the parametric analysis, requires judgements to be made about which specific tidal barrages have the greatest chance of being developed, and on what time scale. For the purposes of the MARKAL energy system modelling, a subset of the Accessible Resource, including the cheapest barrages, was used as the Maximum Practicable Resource input. In the event, the MARKAL analysis estimated no uptake of tidal power under any scenario.

G. Constraints and Opportunities

Technological / System

Tidal energy is highly predictable in terms of both output and timing. It is, however, intermittent, producing ebb generation only once per tide. The annual plant load factor is typically rather low, at about 23%, providing little firm capacity. (The statistical firm capacity contribution of the Severn Barrage has previously been estimated at about 15% of its installed capacity). Thus, the principal effect of building a tidal power station would be to save fuel in existing thermal power stations, rather than to reduce the need for other power stations.

Output would not normally need to be re-timed before being injected into the electricity distribution network. When the barrage is generating, the output of other thermal stations would be turned down and fuel saved. However, some local reinforcement of the network might be needed. For the very large Severn Barrage, the cost of this reinforcement might be more substantial and increase total costs by around 10%.

Over the past three decades much research and development has been undertaken world-wide to reduce the capital costs and/or construction time, and to increase the energy output or its value. There have been substantial improvements in construction methods, machinery and operational techniques and tidal energy technology has moved a long way to maturity. However, although further improvements in generic engineering technology cannot be ruled out, the potential for further cost reductions and improvement in performance is believed to be limited. Continued technical improvement is not expected to reduce the unit cost of tidal energy by more than about 10 to 20%.

Institutional

Non-energy benefits can arise from the building of a barrage. However, these regional benefits are difficult to quantify in money terms and it is not clear how they might accrue to the developer.

Tidal energy projects are capital intensive. However, the cost of capital is affected by risk, as perceived by the lender or investor. For tidal energy, the perception of risk may still be somewhat high because the technology has not yet been demonstrated in the UK and confidence obtained in cost estimation. Some environmental uncertainties also remain. Until tidal energy technology has been more widely established through successful demonstration or commercial operation, the risk factor could tend to raise the cost of commercial capital for such ventures. In view of the scale of finance required and risks involved, the largest barrage projects could require government support in some way, if they are to succeed.

The UK has all the necessary skills and expertise and has, or could construct, all the facilities necessary to build even the largest tidal energy barrage. However, for a very large scheme, such as the Severn Barrage, a large number of turbine generators would be required within a short period. Additional factories would need to be built if importation of some plant was to be avoided.

There could be export opportunities for UK consultants in tidal project design and assessment. Possible markets include Canada, France, CIS, Korea, India, Australia and China.

H. Prospects Modelling Results

Tidal energy resource cost data have been incorporated into the MARKAL model to provide a broad indication of the extent to which the technology is likely to be deployed. Even under the scenario with the highest costs of energy (Heightened Environmental Concern at an 8% discount rate) the model predicts that tidal energy will not be deployed within the next 30 years. The resource-cost data from site-specific studies, particularly the large scale schemes, are believed to be robust with little prospect for major cost reductions. Therefore, despite the relative maturity of the technology, full-scale development even at a small-scale site is very unlikely without specific inducements. To secure private sector finance under current investment criteria, substantial support from the NFFO or a comparable scheme would be necessary, possibly with additional support from government.

II. Programme Review

A. United Kingdom

Tidal energy engineering technology is generally regarded as relatively well developed. However, the technology has not yet been exploited in the UK despite its good technical potential, with the west coast of England and Wales having particularly high tides and suitable sites.

Much of the work undertaken in the UK during the last decade has comprised feasibility or development studies on major sitespecific barrage projects, such as those on the Severn, Mersey, Wyre, Conwy and Duddon. This work has been undertaken jointly by industry, the former Department of Energy and the Department of Trade and Industry through the Government's Renewable Energy R&D Programme. For the Severn Barrage, a consortium comprising six major UK civil and electrical engineering companies was specifically set up to research and promote the project. In the case of the Mersey, a new company was formed: The Mersey Barrage Company.

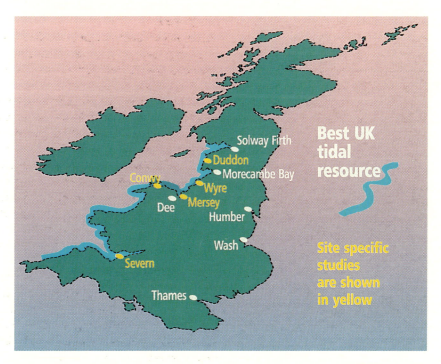

Figure 9: Potential tidal barrage sites.

The cost of electricity from a tidal power station is dominated by the initial capital cost and interest during construction. To try to reduce this energy cost, generic studies have been undertaken to identify alternative methods of construction which might be cheaper and quicker.

There are also some uncertainties about the environmental effects of tidal energy. Generic studies have therefore been commissioned to improve our understanding of estuarine processes and develop improved methods of predicting any likely change.

The Government's Programme

The main objective of the Government's programme has been to reduce uncertainty on costs, technical performance and effects on both the region and the environment to the point where it would be possible to make decisions on whether or not to plan for construction of specific barrage projects. The work being undertaken to meet this objective included studies on:

- the Severn Barrage;

- the Mersey Barrage;

- small-scale tidal energy;

- generic engineering issues;

- generic environmental issues.

Since the publication in June 1988 of Energy Paper 55, 'Renewable Energy in the UK: The Way Forward', substantial progress has been made.

Severn Barrage

A major (£4.2 M) development study was completed and reported in 1989. This confirmed that it would be technically feasible to construct a 16 km barrage across the Severn Estuary in the region of Cardiff to Weston-Super-Mare at a cost of £8.3 billions at 1988 prices (approximately £10.8 billion at 1992 prices), yielding 17 TWh/year from an installed capacity of 8,640 MW. The cost of electricity would be strongly dependent on the discount rate for capital employed. The study also included much work to identify the regional benefits and environmental effects of the scheme. Work on organisation and financing of the project was deferred, pending completion of privatisation of the Electricity Supply Industry.

Public consultation on the report provided a spectrum of views which, on balance, were supportive of the project. The most favourable responses were received from industry, commerce, regional and leisure groups, followed by local and public authorities which tended to be neutral, with more critical comment being received from environmental and ecological groups.

The results of the development project and consultation exercise provided confidence to continue work in three areas:

- further environmental and energy capture studies;

- further regional studies looking particularly at the likely implications of the scheme for the ports and local authorities;

- organisation and financing.

The environmental work and the regional studies have been completed and the study on possible financing mechanisms has been defined.

Mersey Barrage

Three stages of the Mersey Barrage feasibility study have also been successfully completed and reported, bringing the total value of work to date to £7.2 M. A preferred location for the barrage has been selected and

Figure 10: Proposed alignment of the Mersey tidal power barrage.

both energy studies and engineering designs are well advanced. The latest work has concentrated on the accommodation of shipping and on clarifying environmental issues, especially those related to sediments, water quality and birds.

Work to date indicates that the barrage might cost close to £900 M at 1992 prices and take nearly five years to construct, yielding 1.45 TWh/year of electricity from some 700 MW of installed capacity. In 1991 the Mersey Barrage Company (MBC) proposed that the project might be financed by an electricity price of 6.75p/kWh over 25 years, of which 2p/kWh at zero discount rate would be received in advance during the construction period and 4.75p/kWh would be received at the time of electricity supply. Government noted that the 2p/kWh advance payment is equivalent to an interest free lump sum of close to £700 M and that MBC had proposed that this be met in full by a Fossil Fuel Levy under a NFFO contract, and that in addition the 4.75p/kWh subsequent payment would also require NFFO support. On the basis of this financing proposal, Government Ministers considered the project to be too expensive and declined to support further development of the project.

Small Scale Barrages

Initial pre-feasibility studies for small scale schemes on the Conwy (33 MW) and Wyre (64 MW) have also been completed, and a similar study for a barrage on the Duddon (100 MW) is approaching completion.

A further stage of work was undertaken on the Conwy Barrage to optimise its engineering design, and showed that its capital cost could be reduced by 11%. However, this was not sufficient to make the project economically attractive, and further work on it is not anticipated.

The Wyre results are the best so far for a small barrage and indicate that it might produce 131 GWh/year of electricity at a cost of around 6.7p/kWh at 8% discount rate (1992 prices.) Although this is similar to the cost of electricity from the best of the larger barrages, Ministers have declined to support further development of the project.

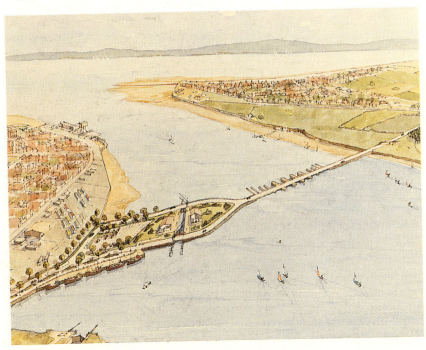

Figure 11: Artist's impression of a tidal energy barrage across the Wyre Estuary.

Generic Studies

Government has also continued to support a range of generic studies. These have underpinned the commercial site-specific barrage studies, by exploring possible methods of lowering costs and reducing technical and environmental uncertainties, and have mostly been fully funded by government. They have comprised the following.

- Generic engineering studies to develop alternative methods for constructing tidal energy schemes, modelling their essential parameters and forecasting their costs and benefits. The possible use of steel caissons and in-situ construction using concrete diaphragm walling built within a temporary sand island have been considered. The dynamics of caisson placement have been researched. The merits of composite materials, such as steel/concrete/steel sandwiches, are also now being investigated. The case for extracting the kinetic energy from tidal streams, by placing turbines in the free-flowing currents without a barrage, has also been examined. Although the UK technical potential for tidal stream energy, approximately 58 TWh/year, is similar to that for tidal barrage energy, costs are high, being in the range 10 to 60p/kWh at 8% discount rate for larger machines, and higher for machines rated at under 100 kW. At these costs exploitation would seem unlikely within the foreseeable future.

- Generic environmental studies have been undertaken to clarify understanding of estuarial processes and develop improved methods of forecasting the changes which might occur on construction of a barrage. Work is continuing to improve the prediction of possible changes in sediment stability, water quality, invertebrate populations, wading bird and wildfowl populations, and the extent of saltmarsh. The potential effects on fish are being investigated and improved means to ensure their safe passage across the barrage are being evaluated.

B. Other Countries

In some other countries, work has also been carried out to investigate the potential for tidal energy and assess the opportunities afforded by specific sites. Also, following the construction of prototypes and demonstration plant in France, Canada, USSR and China (see table 1), work has continued to develop the technology.

The Bay of Fundy, Canada, has the highest tides in the world and a variety of potential sites which could be exploited. Of these, the Cumberland Basin is not considered competitive at present and the Minas Basin, although economically more attractive with unit costs about 30% lower, is thought to have unacceptable impacts on the US coast of Maine. Work completed recently has included investigation of the technical possibilities and costs of re-timing output by various means, including compressed air storage, to facilitate the possibility of exports to the US. Environmental studies have been undertaken on sediment stability and also on fish diversion techniques to limit damage by the turbine. In addition, performance of the Annapolis rim-generator turbine has been monitored and its reliability established.

The former USSR has been active in assessing its large potential, both in the White Sea and in the Sea of Okhotsk. In the latter area, a feasibility study for a major scheme at Tugursk has recently commenced. Provisional plans call for a total installed capacity of 1,000 MW by 2005; 5,000 MW by 2010 and 10,000 MW by 2015.

Garolim in Korea and the Gulf of Kutch site in India have been evaluated. Mexico is also investigating a site in the Colorado Estuary. China is understood to be conducting feasibility studies at two further sites; its south-east coast in particular is considered to have substantial potential. Australia is reinvestigating the possibilities of tidal energy sites in the Secure Bay area, but the remoteness of potential markets for the power is a major constraint.

Interest in some sites has lapsed because previous studies found costs too high. Among these are sites in Alaska, Passamaquoddy in the Bay of Fundy, San Jose in Argentina and the north-east coast of Brazil.

Other areas with large energy potential but not the focus of present interest include the Mont St Michel Bay in France and Ungava Bay in Northern Canada. The former has been studied in depth but financial constraints make development unlikely; the latter site is too remote.

Through ETSU a careful watching brief on overseas developments is being maintained. Some limited collaboration is also taking place, including information exchange with Canada, France and the CIS, particularly on environmental aspects. On the experimental front, Canadian experts are assisting UK generic studies on sediment stability, while France has allowed the UK access to its La Rance scheme for studies on the safety of fish. Outside the UK Government's programme, UK consultants have been active in assisting with the Korean and Australian feasibility studies described above.

Tidal Power

Wave Energy

I. Technology Review

A. Technology Status

Waves in the oceans are created by the interaction of winds with the surface of the sea. Winds are themselves caused by a combination of the sun's heat and the rotation of the earth. Ocean waves contain large amounts of energy stored in the velocity of the water particles and in the height of the mass of sea water in a wave front above the mean level of the sea. The amount of the wind energy that can be transferred to the surface of the ocean to create the waves depends upon the wind speed, the distance over which it interacts known as the fetch and the duration for which it blows over the water. Because of the direction of the prevailing winds and the size of the Atlantic Ocean, the UK and North Western Europe have one of the largest wave energy resources in the world. Figure 1 shows the global distribution of wave energy.

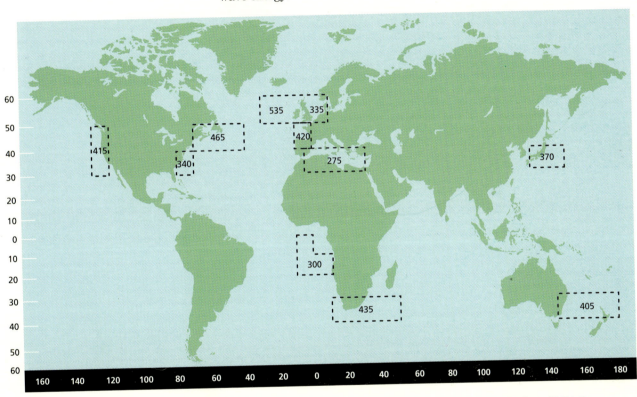

Figure 1: Global wave energy levels (as the annual energy per metre of wave front (MWh/m)).

The principal ways of extracting energy from waves rely either separately or jointly on the surge, heave and pitch of the waves as shown in figure 2. The frequency of arrival of ocean waves is low (a few per minute). The conversion mechanism must therefore produce a higher frequency rotation to generate electricity - which is a convenient energy transfer medium. This can be done by hydraulic pumps or pneumatic bags/chambers driving higher speed turbines and generators. Such conversion mechanisms have been tested by the construction of many varieties of experimental laboratory models and by some small-scale devices in the open sea or large lochs.

The UK Department of Energy (DEn) funded extensive research into wave energy during the period 1974 to 1983 under its wave energy programme, spending over £17 M in money of the day. More than 300 different concepts of machine for extracting the energy from the waves by converting it to

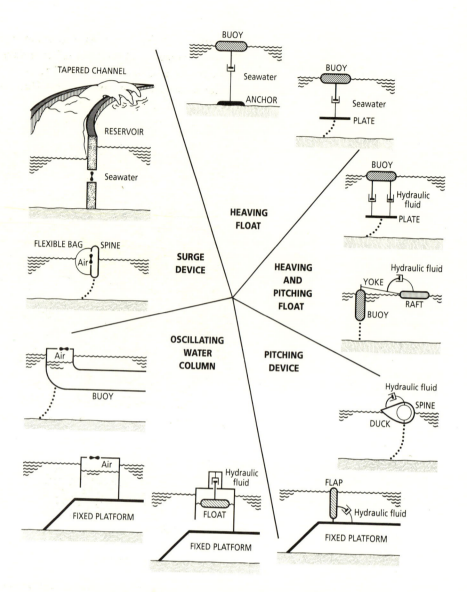

Figure 2: Categories of wave energy devices *(courtesy Hagermann and Heller)*.

mechanical motion and subsequently to electrical energy were considered. Eight of these ideas were taken to the design stage of a 2 GW large-scale offshore system. Since 1985, the Department of Trade and Industry has concentrated its R&D effort on the most promising shoreline device. This is an oscillating water column (OWC) with a novel air turbine named after its inventor Professor Wells. The RD&D to develop a shoreline gully device of 75 kW capacity has been carried out by the Queen's University of Belfast. This device has been commissioned on Islay in the Hebrides and is currently producing electricity for the grid.

Currently there are two types of device in operation. One type is a shoreline or caisson breakwater oscillating water column driving a pneumatic Wells turbine (in 10 to 25 m depth of water).

A Wells turbine rotates in the same direction irrespective of the direction of the air flowing over it. It is ideally suited for connection to an oscillating water column or flexible bag that pushes air out and then draws it in with each rise and fall of the wave. The UK shoreline device on Islay is a partially buried rectangular concrete box on a rocky shoreline with an enclosed turbine. The exposed nature of such sites dictates that the electricity and control cables are buried alongside a small control building

Figure 3: Islay shoreline wave energy device.

and also that the turbines and generators are enclosed for protection. Figure 3 shows a photograph of this device and figure 4 shows a cross-sectional view of the chamber and turbine. Figure 5 shows the construction sequence proposed by the Queen's University of Belfast for the artificial gully oscillating water column.

Figure 4: Artist's impression of a section through the Islay wave energy device.

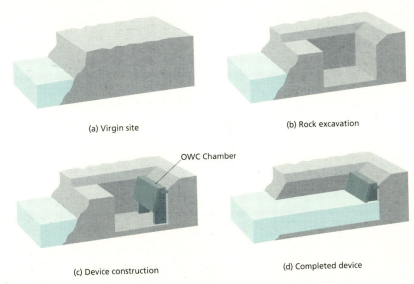

Figure 5: Construction sequence proposed by the Queen's University of Belfast for the artificial gully oscillating water column. *(Courtesy the Queen's University of Belfast.)*

A second type is a tapered channel device that was developed in Norway. In this type, incoming waves travel up a tapering channel, overflow and fill a higher level reservoir. The enclosed water then drives a Kaplan hydroelectric turbine as it returns to the sea. This device has been undergoing reconstruction since 1991 and is intended to be recommissioned in 1994.

The life expectancy of the civil engineering components of a shoreline device is anticipated to be 30 to 40 years - although turbines operating in marine conditions would have shorter lives (approximately 10 years). Decommissioning of shoreline devices has not yet been addressed because the structures built so far are fairly substantial concrete enclosures. However, the decommissioning problem is similar to that for North Sea

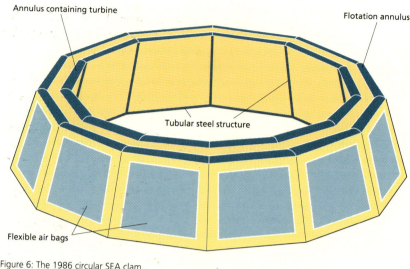

Annulus containing turbine

Flotation annulus

Tubular steel structure

Flexible air bags

Figure 6: The 1986 circular SEA clam.

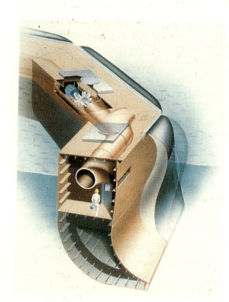

Figure 7: Cross section in steel of the 1991 circular clam.

platform disposal which is currently being investigated by both government and the oil industry.

A floating offshore device known as the circular clam has been developed by Sea Energy Associates (SEA) and Coventry University. The design comprises a floating twelve-sided hollow ring. Each of the twelve sides has a flexible membrane that inflates and deflates with the incoming wave action. The air passes via a central circular manifold through Wells turbines which drive electrical generators.

Figure 6 shows a schematic drawing of the original clam design with the air bags and the manifold or annulus. Figure 7 shows the steel hulled version of the 1991 design that is some 60 m in diameter. The 12.5 MW scheme outlined in table 5 is for five clams moored off South Uist.

B. Market Status

A summary of the current status of the technology is shown in table 1.

Current status	Shoreline:	Experimental prototype (UK) Commercial device (Norway)
	Nearshore:	Experimental prototypes (Japan & India)
	Offshore:	R&D in wave tanks
UK installed capacity	Shoreline:	75 kW prototype
	Nearshore:	None
	Offshore:	None
World installed capacity	Shoreline:	~500 kW
	Nearshore:	~150 kW
	Offshore:	None
UK industry	None	

Table 1: Status of technology.

Offshore devices for deep water are still at the R&D stage.

Shoreline or nearshore sea bed mounted wave energy devices are currently at the development stage. A possible exception is the Norwegian tapered channel (Tapchan) device, a 1 MW version of which is being considered for islands in the South Pacific. The tapered channel device is likely to be restricted to regions of low (1.5 m) tidal range for technical reasons. It is not suitable therefore for most of the UK except perhaps the Shetlands. Apart from the developers of this device there are at present no commercial wave power device manufacturers.

The market for shoreline devices centres on islands where grid connection is expensive or unreliable and where diesel powered generation is consequently the main competitor. Caisson breakwaters may have a market in countries requiring new or refurbished coastal defences and/or harbours and where the wave energy extraction plant is a marginal addition to the costs. In such cases however, when a high average incident wave energy is necessary, a few intermittent high energy storms per year may not justify the cost of installing a wave energy device. The technology has been deployed by China, Denmark, India, Japan, Norway, Spain, Sweden and the UK. In the UK and Norway there may be some economic resource. Japan and India have lower wave energy resources but have need for further coastal protection and could take advantage of marginal breakwater construction costs.

No support has been given to wave power under the Non-Fossil Fuel Obligation since the technology is still essentially in the R&D stage.

C. Resource

The UK wave energy resource can be subdivided into two categories (shoreline and offshore). The Accessible Resource available at a cost of less than 10p/kWh (1992 prices) has been evaluated as part of the 1992 Wave Energy Review and is shown in table 2.

Discount Rate	England & Wales	Scotland	Northern Ireland	UK
8% Shoreline	0.01 TWh/year	0.39 TWh/year	0	0.4 TWh/year*
8% Offshore	0	~ 0.03 TWh/year	0	~ 0.03 TWh/year
15% Shoreline	0.01 TWh/year	0.29 TWh/year	0	0.3 TWh/year*
15% Offshore	0	0	0	0

Table 2: Accessible wave energy resource for the UK at less than 10p/kWh (This data assumes the successful completion of all outstanding R&D on these devices.)
Note: *The shoreline resource is very dependent upon the details of the local shoreline.

Figure 8 shows the principal offshore wave energy resource level in kW/m of wave front. The shadowing effects of Ireland on the western coast of England and Wales can be seen in this figure.

There are two constraints that might prevent the full utilisation of the offshore resource if the technology were available. The first is the requirement for navigational and fishing controls around and among the large devices. The second is the need for suitable high power grid connections from the offshore devices to transmit the electricity to the National Grid for use in central England. For shoreline wave energy extraction there could be different problems. In remote locations and on islands there would be the need for land based access and possible road construction. For all sites there would be a desire to conserve the natural rocky coastline and the associated sea shore. Finally, on mainland sites there could be the need for a peripheral coastal grid connection that may be visually intrusive.

Figure 9 shows regions where energy levels have been evaluated from measured offshore wave spectra and charted seabed contours. Frictional modelling programmes have been used to compute the available energy in a wave front at the 25 m depth contour and at the shoreline (taken to be 10 m depth). In principle the different wave energy resources are not mutually exclusive. Deployment of large offshore devices around the north western UK coastline would reduce the available wave energy at the shoreline.

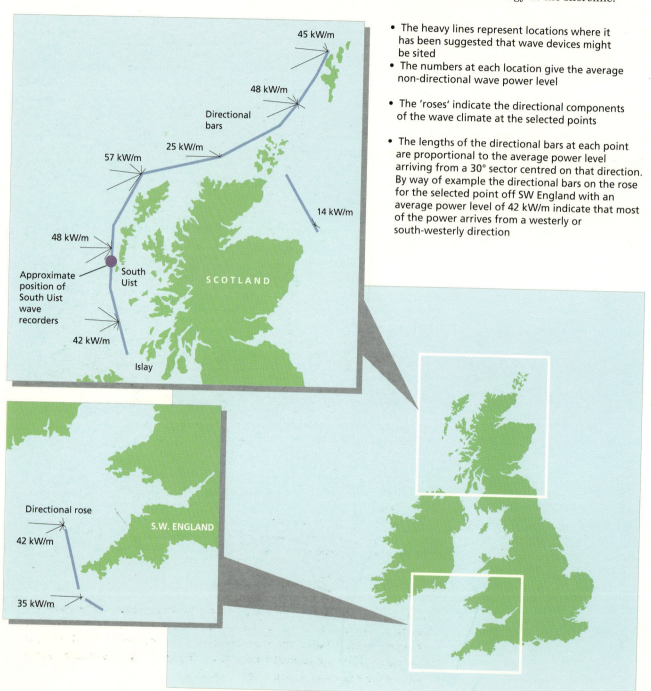

- The heavy lines represent locations where it has been suggested that wave devices might be sited
- The numbers at each location give the average non-directional wave power level
- The 'roses' indicate the directional components of the wave climate at the selected points
- The lengths of the directional bars at each point are proportional to the average power level arriving from a 30° sector centred on that direction. By way of example the directional bars on the rose for the selected point off SW England with an average power level of 42 kW/m indicate that most of the power arrives from a westerly or south-westerly direction

Figure 8: Variation of wave power with location around the UK coast.

D. Environmental Aspects

Wave energy devices produce no gaseous, liquid or solid emissions and hence, in normal operation, wave energy is a non-polluting energy source. Table 3 shows the environmental burdens associated with the technology. Any significant usage of land would be restricted to the shoreline technology and the specific area of land required could be of the order of 200 sq.m per MW of capacity. However, the siting of shoreline plants would need to be on exposed rocky coasts below the high water mark and such land is regarded as environmentally and

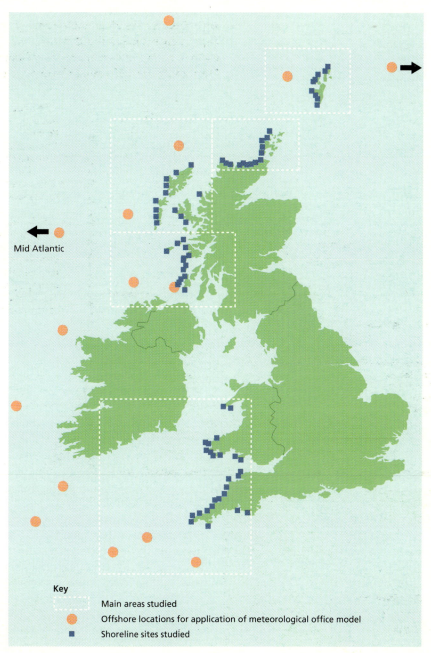

Key

    Main areas studied

    Offshore locations for application of meteorological office model

    Shoreline sites studied

Mid Atlantic

Figure 9: UK wave energy sites.

aesthetically valuable. The construction of access roads, the need for distribution cables along the coast and the construction of visually intrusive control buildings may be considered unwelcome. Finally, if substantial off-shore capacity were to be deployed, there would be some land use for substations and grid connection.

Nearshore devices, either floating or seabed mounted caissons, require little or no land but, depending upon the distance from shore, they can have similar environmental impacts to equivalent land-based installations. One class of device proposed would be mounted on the seabed near steep cliffs to utilise the energy available in waves in a water depth of 10 to 25 m. These devices would however be placed sufficiently close to the shore to enable the power cables to reach the land by a catenary (rather than being submerged on the seabed) and allow access for the maintenance crews by walkway or cable car (rather than by boat).

Environmental burden	Specification
Gaseous emissions	None
Ionising radiation	None
Wastes	None
Thermal emissions	None
Amenity/Comfort:	
Noise	Shoreline devices using air turbines: There are no experimental data at present but noise levels may be similar to wind turbines although enclosed and at lower elevation (3 to 5 m)
Visual intrusion	Shoreline: some impact, possible mean elevation of plant of order 5 m
Use of Wilderness area	Shoreline: significant impact expected on rocky coasts of islands during construction and from surface grid connection
Other	Nearshore and offshore: constraints on navigation and fishing
	Large scale offshore: attenuation of the ocean waves incident upon the UK coastline

Table 3: Environmental burdens associated with wave energy.

Noise from shoreline and nearshore devices may result from the air turbine used with an oscillating water column. The current phase of work on the Islay shoreline device will address the volume, frequency distribution and methods of mitigation of any turbine noise. No large-scale offshore devices have yet been constructed but the importance of the following environmental effects - some of which may be considered beneficial and others detrimental - would depend upon the choice of site:

- new areas of sheltered water;

- attraction for fish, sea-birds, seals and seaweed;

- low frequency noise under water which may affect seals, whales etc.

- possible effects on tidal currents.

At coastal locations, an effective wave power station would modify the local wave climate. Offshore floating devices of low freeboard, unlike seabed-mounted devices, would probably have a minimal effect on the coastline. Any reduction of the wave energy reaching the shore and the effects on shallow sub-tidal areas could change the density and species of resident organisms.

Wave energy devices could be a hazard to shipping due to their low freeboard rendering them relatively invisible either by sight or by radar. This could be minimised by marking the devices with lights and transponders and by providing navigation channels in the device arrays. Wave energy devices drifting free as a result of mooring failure would be a hazard, not only to ships but also to coasts and harbours landward of the station. Repair and rescue resources would need to be deployed sufficiently rapidly to retrieve them before they reached land. Such techniques and procedures have been demonstrated in the offshore oil industry. They have had to be developed because of the commercial nature of offshore oil and gas extraction operations.

Table 4 shows the potential emissions savings resulting from the use of wave energy.

Emission	Displaced emissions (grammes of oxide per kWh electrical)	Annual savings per typical 1 MW$_e$ scheme generating 37 GWh/year (tonnes of oxide)
Carbon dioxide	734	1,100
Sulphur dioxide	10	15
Nitrogen oxides	3.4	5.1

Table 4: Emission savings relative to generation from the average UK plant mix in 1990.

For the calculations in table 4, it is assumed that each kWh generated by wave power displaces a kWh generated by the average plant mix in the UK generation system (for the year 1990) and thereby saves its associated emissions.

E. Economics

Table 5 shows the expected costs and performance data for two representative wave energy devices studied in the 1992 Wave Energy Review. These are a 1 MW shoreline gully OWC and a 12.5 MW scheme of 5 circular clams for offshore deployment. The data are in 1992 values with an inflation factor of 1.09 on the 1990 data published in the Review. In order to address the relatively undeveloped nature of wave energy, the Wave Energy Review gave considerable allowance for the potential benefits of R&D on the practicality, performance and costs of the wave energy devices considered. In keeping with the forward nature of the Wave Energy Review, this Assessment adopts the electricity generation costs predicted for the representative wave energy devices allowing for the full benefit of the successful completion of all identified future R&D.

Plant type		1 MW Shoreline OWC	12.5 MW scheme of 5 1991 Circular Clams
Capital cost		1.53 £k/kW	2.6 to 3.0 £k/kW
	Structure	£496,000	£8.8 M to £1.1 M
	Mechanical & engineering plant	£216,000	£3.4 M
	Transmission	£206,000	£3.2 M
	Miscellaneous	£390,000	
	Project management & contingency	£225,000	£1.5 M to £1.8 M
Annual recurring costs	Operation & maintenance	44 £/kW	66 £/kW
Performance data	Plant availability	97%	96%
	Average load factor	25%	24%
	Resource availability	26%	25%
	Type of operation	Intermittent but predictable	
Other	Construction time	1 year	2 years
	Lifetime	25 years	25 years

Table 5: Cost and performance data for a 1 MW shoreline oscillating water column and a 12.5 MW scheme of five circular clams (These data assume the successful completion of all outstanding R&D on these devices).

Season	Day/Night
Spring/Autumn (April & September)	20%
Summer (May to August)	10%
Winter (October to March)	70%

Table 6: Annual distribution of energy generated by a typical wave device.

F. Resource-Cost Curves

Figure 10 shows the Accessible Resource for offshore wave energy. This is based on reference systems of several circular clams operating off South Uist. The need to synchronise the generation with the National Grid, lower wave power levels at other sites and the cost of grid connections for large

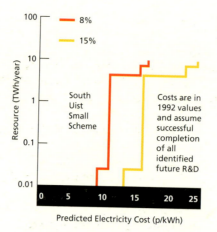
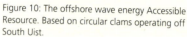

Figure 10: The offshore wave energy Accessible Resource. Based on circular clams operating off South Uist.

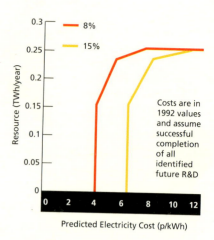

Figure 11: Maximum Practicable shoreline resource; year 2025, 1992 prices

schemes would all increase the cost of subsequent devices. This causes the resource cost curve to rise above the 10p/kWh level. Costs are in 1992 values and assume successful completion of all outstanding R&D.

The Accessible Resource for shoreline wave energy is estimated to be 1.7 TWh/year.

The Maximum Practicable Resource, for both offshore and shoreline wave energy, takes into account visual intrusion, ecological and environmental factors and can only be a speculative estimate at this stage.

The Maximum Practicable Resource for offshore wave energy by 2025, based on current designs and the assumptions in the Wave Review, is estimated to be less than 10 TWh/year.

Figure 11 shows an estimate of the Maximum Practicable Resource for shoreline wave energy for the coastal regions shown in figure 9, as calculated by the Queen's University of Belfast for the Wave Energy Review. This figure takes account of transmission losses and is constrained by institutional factors such as the limited market for electricity on islands.

G. Constraints and Opportunities

Technological / System

There are technical constraints that apply to both categories of wave energy device (shoreline and offshore). They are as follows.

- It must be demonstrated to industry and potential developers that a device can be designed, constructed and deployed successfully within the costs and time scale proposed. At present, apart from the Islay shoreline device, there is little available data and the construction and emplacement costs and the risks for offshore devices are perceived to be high.

- The technical performance and reliability of wave energy devices over many years of varying sea conditions (including storms) must be demonstrated.

- For nearshore and offshore devices there are additional constraints.

- The applicability of conventional offshore mooring technology has to be established for floating devices, and seabed emplacement techniques for fixed devices must be developed and demonstrated. These technologies have been developed for shipping, barges and offshore oil platforms but have, of necessity, been designed to have low interaction with the waves – the opposite to that required for wave energy extraction devices.

Institutional

If wave energy technology was developed to the commercial exploitation stage other constraints of an institutional nature would then need to be faced. They would include the following.

- Shoreline sites would require planning permission.

- The normal operation of floating offshore devices and the period of emplacement of seabed mounted devices would require the necessary third party insurance. Lloyd's or any other insuring organisation would have to include commercial devices in the appropriate regulations for construction of offshore structures.

- The rights to navigation channels for fishing, commercial and naval shipping would require consideration.

- There could be multiple ownership and access rights to shoreline and nearshore sites. Both the landowners and the Crown (for land below the high water mark) would own the rights. This could affect the device construction, access and connection to the electricity grid.

H. Prospects Modelling Results

Table 7 shows the predicted uptake for shoreline wave energy from the MARKAL model. As can be readily seen, there is only an uptake under the Heightened Environmental Concern scenarios. The predicted uptake is approximately 20 MW expressed as a mean annual capacity or 80 MW of installed capacity.

Scenario	Discount rate	Contribution, TWh/year			
		2000	2005	2010	2025
HOP	8%	0	0	0	0
	15%	0	0	0	0
	Survey (10%)	0	0	0	0
CSS	8%	0	0	0	0
	15%	0	0	0	0
	Survey (10%)	0	0	0	0
LOP	8%	0	0	0	0
	15%	0	0	0	0
	Survey (10%)	0	0	0	0
HECa	8%	0.16	0.16	0.16	0.16
	15%	0	0	0	0
	Survey (10%)	0	0	0	0
HECb	8%	0.16	0.16	0.16	0.16
	15%	0	0	0	0
	Survey (10%)	0	0	0	0

Table 7: Shoreline wave potential contribution under all scenarios.

The environmental sensitivity studies and Shifting Sands scenario do not predict any change from the base case shoreline wave uptakes.

Given the early stage of development of shoreline wave energy devices, the development of the shoreline resource by the year 2000 may be considered optimistic.

Data describing estimates for the future costs and resources for offshore wave energy devices, assuming the successful completion of an extensive R,D&D programme, consistent with that identified in the Wave Energy Reveiw, were supplied to the model but no deployment under any scenario or discount rate occurred.

II. Programme Review

A. United Kingdom

Wave energy technology is still essentially at the research stage. Work on the technology is confined principally to University research teams, some Government laboratories and engineering consultancies. A wide and diverse range of small-scale experimental devices exist but few have prospects of leading to a practicable and commercially viable technology. Industry's perception is that the offshore environment is difficult to deal with and there are many intractable problems concerning the survival and subsequent maintenance of offshore devices and their connections to the grid.

The Government's Programme

Since the publication of Energy Paper 55, the Government's programme has continued to provide support for R&D on shoreline wave energy and to assess devices having prospects of generating electricity at an economic cost. A comprehensive technical and economic review of wave energy in the UK has been carried out by the Department of Trade and Industry. This included specific studies by consulting engineers on the availability, performance and consequent economics of a range of nearshore and offshore devices. These independent consultants worked in co-operation with the design teams of the principal wave energy devices.

For the shoreline technology, the emplacement, grid connection and operational problems are mitigated to some degree. Also for islands and remote locations the competing technology is usually relatively expensive diesel generation. A 75 kW oscillating water column installation, designed and commissioned by the Queen's University of Belfast for the Department of Trade and Industry, is now operating on the island of Islay in the Inner Hebrides. This machine uses a natural rock gully and is providing electricity to the local grid, which in turn is connected to the National Grid. A three year programme of monitoring and optimisation is in progress. The objective is to identify the technical, operational and environmental problems of such devices and define satisfactory solutions to bring the technology to a level where the construction of a megawatt-scale demonstration plant could be considered. In parallel with this work, a detailed economic assessment of the UK shoreline and nearshore wave energy resource has been carried out, modelling the shoreline wave energy levels and quantifying the infrastructure constraints on all potential sites.

The main industrial involvement in the shoreline programme has been with Hydro-Electric plc (formerly North of Scotland Hydroelectric Board) through their provision of grid connections. Continued Government involvement is likely to be necessary in the development of shoreline devices if it is to be taken to the demonstration stage, and the market for it determined. The Department of Trade and Industry provides the principal funding for the UK programme. Some complementary funding is provided by the Marine Technology Directorate of the Science and Engineering Research Council to support work on wave loadings on the Islay device.

B. Other Countries

The main initiatives abroad have been taken by China, Denmark, India, Japan, Norway, Portugal, Spain and Sweden. The Japanese and Indian programmes have concentrated on oscillating water column devices using Wells turbines, constructed in breakwater caissons as part of seawall or harbour defences. This mitigates to some extent the civil engineering and emplacement costs. However, the natural wave climate of these countries is less energetic than that of the Western European coasts. In the past Norway developed an oscillating water column machine but this was destroyed in a severe storm. Norway has also operated a tapered channel device feeding a lagoon or reservoir (known as a Tapchan). It is planned to modify and re-commission this device in the summer of 1994. It is claimed that it is close to being commercially viable and larger installations in the Pacific are being planned. More recently a 110 kW pilot floating wave power device was tested off the west coast of Sweden. This consists of a steel platform containing a sloping ramp which gathers incoming waves into a raised internal basin. The water flows from this basin back into the sea through low-head turbines. In these respects it is similar to an offshore Tapchan. However, the device is not sensitive to tidal range and, by varying its ballast, the device can be 'tuned' to different wave heights. Discussions are currently underway for development of a 1 MW version of this device. Portugal is developing a shoreline device in the Azores for the local electricity utility. There has been no major overseas development of large-scale offshore devices other than earlier deeper water experiments by Denmark and Sweden; these used floating buoys which operated either seabed or hose pumps.

A programme of R&D on wave energy has been initiated by DGXII of the CEC. The work so far has concentrated on the following:

- wave studies and development of a resource evaluation methodology;

- a generic technical evaluation study of wave energy converters;

- studies of device fundamentals and hydrodynamics, components and materials, power take-off systems, electrical systems, control and grid interaction;

- a study of the methodology for reliability, costing and environmental impact;

- a site selection study for a 500 kW to 1 MW capacity Shoreline OWC European pilot plant.

A call for proposals in 1993 has addressed work on these topics.

Photovoltaic Systems

I. Technology Review

A. Technology Status

Photovoltaic (PV) materials generate direct current electrical power when exposed to light. The photovoltaic effect was discovered by Becquerel in 1839. To date it has been most notably exploited in space applications for powering satellites where the low power requirements, good solar resource and simplicity of operation have outweighed the high cost of the devices. Terrestrial applications include various consumer goods, lighting and water pumping systems for developing countries and remote areas with no grid supply. R&D and developments in manufacturing techniques over recent years have reduced the cost and improved the performance of PV systems; in response to this a review of the technology was carried out in 1990 under the DTI Renewable Energy Programme. The conclusions of this review were that large-scale centralised generation of electricity for the grid by means of PV is unlikely to be economically attractive in the UK, at least for the foreseeable future. Distributed generation using small-scale systems operated by consumers and remote site autonomous systems are likely to be more attractive options. These could become economic in the longer term. The review concluded that under favourable scenarios of the future certain PV applications, in particular systems integrated into buildings, could offer a generation cost competitive with other renewable sources.

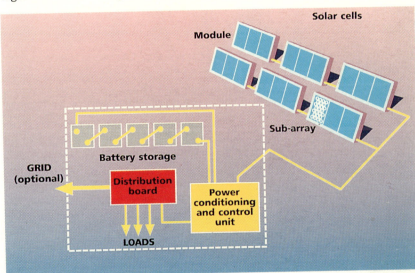

Figure 1: Schematic of a typical photovoltaic system.

An optimised building integrated PV system for a house might have a specification and performance similar to that set out below, using current generation PV technology:

- area: 66 sq.m, inclined @ local latitude (~50°), oriented south;

- installed capacity: 3.3 kW (peak);

- average annual output: 2.6 to 3.3 MWh/year (depending on location);

- estimated system cost £5,500 (allowing for savings from displaced roofing materials);

- estimated electricity cost (8% discount rate and 25 year lifetime) 15 to 20 p/kWh.

For comparison, the current average domestic electricity demand in the UK is around 4.6 MWh/year per household. It is anticipated that future developments in PV technology could increase the scale of output to between 5 and 6.5 MWh/year from a system of this scale. In principle, therefore, a significant fraction of the domestic electricity demand could be met by PV systems. It is unlikely, however, that the load profile would match the supply curve so storage or export to the grid would be required. Non-domestic buildings could support larger systems and might be able to make better use of PV power since most of the energy is required during the day time.

Photovoltaic materials are usually solid-state semiconductors which generate an electrical potential when exposed to light. The voltage generated is a function of the band gap structure of the semiconductor used and the physical form of the material – either crystalline or amorphous. It is typically around 1 volt. The PV material and its electrical contacts are usually referred to as a cell. The collection and encapsulation of cells into a standard sealed unit of given voltage and current is termed a module.

The main semiconductor materials currently used are:

- mono-crystalline silicon;

- poly-crystalline silicon;

- amorphous silicon.

In some specialised applications, for example in space technology, gallium arsenide has been used in place of silicon because of its higher efficiency. However, the higher costs of this and other exotic materials currently preclude their widespread application.

Some other materials are currently under development, particularly for thin film type applications. These may lead to lower mass production costs coupled with high efficiency. The main front runners are cadmium, telluride and copper indium diselenide. In addition, new cell structures are under investigation including multi-junction devices which show promise of higher stable efficiencies.

The performance of PV systems can be characterised in terms of cost and efficiency by their cost per Watt of power generated. Because the generation output is a function of incident irradiation, spectral content and module temperature it is usual to normalise the output to that obtained under 1,000W/sq.m at 25°C under standard atmospheric conditions. This is referred to as the peak power output since most terrestrial applications would have a lower insolation level than this. Thus modules are normally characterised by their Watt peak performance (Wp) and their unit cost £/Wp.

The economics of PV systems are, to a large degree, determined by this parameter and the light-to-electricity conversion efficiency. The lower the £/Wp the better but the lower the £/Wp the more the system cost is determined by the balance of system (BOS) elements. The £/Wp parameter can be reduced by increasing the efficiency of conversion of light to electric power and by reducing the manufacturing costs of the modules. Higher efficiencies also mean that for the same peak power a smaller area of module is required. This can reduce some of the requirements for the BOS. Advances in both of these areas suggest that PV may become economically attractive in the future, despite the relatively modest levels of solar radiation experienced in the UK.

The two main manufacturing processes currently adopted produce crystalline or poly-crystalline silicon cells and amorphous silicon cells respectively. Each process has advantages, the crystalline cells having higher efficiencies but also higher manufacturing costs per square metre. For most current off the shelf modules of either type, the typical cost is in the range £2/Wp to £3/Wp. Over the next decade it is thought that this could fall to £1/Wp or less with new materials and techniques currently under development. Typical efficiencies are in the range 5 to 16% for commercial modules, depending on the material type. Mono-crystalline silicon is at the higher end of the range and amorphous silicon at the lower end.

A PV system would include various other components, depending on the application. There is probably less scope for significant technological developments to reduce the costs of these BOS elements, although costs would reduce through mass production if a significant market developed. However, as the cost of the PV modules decreases, the BOS costs will form an increasing fraction of the total system cost. A particular attraction of building integrated PV systems is that the structural support elements could be part of the building itself. This would reduce the overall cost of the PV system. In addition, the ability to connect to an existing grid supply may remove the need for local energy storage, as well as allowing the export of power to the grid when not needed.

Systems are expected to have a lifetime of at least 25 years with low maintenance requirements. However, as yet, there is little experience of operating PV systems over long periods.

B. Market Status

The worldwide PV manufacturing industry has an estimated annual turnover of about $300 M. Most of the major manufacturers of modules are located in North America, Europe and Japan, but there is a growing number of smaller manufacturers in more recently industrialised countries such as India and China. Many of the major manufacturers have diversified into PV from related industrial or energy activities, such as electronics and oil supply. It is only relatively recently that the market has reached a size large enough to support a sizeable manufacturing base. In 1992 the total manufactured output of PV modules was about 60 MWp worldwide with the last few years showing a year on year growth of about 15%. The continued expansion of this market is crucial if unit production costs are to be reduced. In the short term at least, this expansion will continue to be in the remote power sector market.

In addition to the module manufacturers, there are a large number of end-use system suppliers who design and assemble appliances, such as water pumps and lighting systems, from PV modules and other components such as batteries, charge controllers etc. This activity can be more profitable than PV module manufacturing itself and so several module manufacturers also provide complete systems for target markets.

The current markets for PV systems can be identified as consumer goods, remote site/stand-alone systems and grid connected systems. In the UK it is the consumer goods and small remote power areas that are currently largest and may be expected to grow most rapidly in the short term. There are relatively few examples of professional remote power systems and grid connected applications. At the present time little information is available on the actual size of these markets because of their diversity and small size. An estimate for 1992 indicated the following breakdown:

PV sales in the UK in 1992
- consumer goods   15 kWp
- remote power   25 kWp
- grid connected   less than 5 kWp.

On the world market there is more information available and a more accurate analysis is possible. The market sectors are the same as for the UK but their sizes are much larger.

World PV sales in 1990
- consumer goods   10 MWp
- remote power   34 MWp
- grid connected   7 MWp

The current market status is summarised in table 1.

Current status	Consumer goods: commercial Remote power: commercial/demonstration Grid connect: development/research
UK installed capacity	< 1 MWp (estimated)
World installed capacity	> 300 MWp (estimated)
UK industry	Small number of manufacturers Research community

Table 1: Current market status.

C. Resource

There are two aspects to the resource size for PV systems, firstly the solar energy resource and secondly the area requirements for deployment.

The solar resource is independent of the application and is simply the total solar insolation. In the UK, the available sunlight is split between the direct sunlight component and the diffuse sky component. The split of the total radiation between these varies according to the cloud cover, but on average over a year it is about 50:50. For non-focusing flat plate PV systems, the split of the radiation into direct and diffuse components is not very important but greater levels of direct solar will increase the output attainable. The average W/sq.m falling on a module over a year can be crudely optimised by tilting the module at an angle equal to the local latitude. Fine tuning of the tilt angle to track the varying height of the sun during the year is not generally worthwhile since the small gain in output is offset by the increased costs and complexity of the tracking system. The annual average insolation values for the UK range from 2.2 kWh/sq.m/day in northern Scotland up to 3.0 kWh/sq.m/day in southern England. There is significant variation around this average value due to both seasonal effects and daily weather variations. These insolation levels result in load factors (the fraction of the theoretical maximum annual output of 8.76 kWh/Wp) of 10 to 15 per cent.

The area required for deployment is dependant on the application of the technology. Central generating systems require land area; building integrated systems use the building for support.

Central generation

At present, the economics of PV central generation in the UK are unattractive. Consequently no detailed assessments have been made of the land use aspects and the requirements of such plant. The Accessible Resource is simply the land area remaining after allowing for natural and man-made uses and restrictions, i.e. buildings, roads, lakes, National Parks etc. The Maximum Practicable Resource is substantially less than this for several reasons. The basic infrastructure and services requirements, such as access, spacing to avoid overshading, control rooms, switching gear etc. will limit the packing density of modules per square kilometre of plant. More significantly, other land use requirements (agriculture and recreation for example) will substantially reduce the areas available for PV systems since parallel uses of the land for other purposes will be precluded. Although PV systems are essentially silent in operation there would be considerable visual intrusion on the landscape and this would raise questions of public acceptability. There are schemes for using "waste" land areas, such as motorway verges, for the installation of PV systems; this approach would reduce visual intrusion and avoid competition with other claims on the land.

Building integrated systems

The Accessible Resource for building applications is essentially the areas of roofing and walls of the existing building stock which could be retrofitted with PV systems. Some physical factors would constrain this, such as glazed areas etc. but, except where a retrofit programme is being carried out for other reasons, this is unlikely to prove economic. In new buildings, however, where the PV components can be designed into roof and wall areas displacing building components, and the building structure and module support system can be combined, economic viability shows prospects. A detailed assessment of the UK Accessible Resource from building-mounted systems has been carried out, taking into account shading, unsuitable buildings, architectural requirements and other constraints. This indicates a resource of up to 200 TWh/year, in the short term and up to 360 TWh/year, in the long term (2020). This study included all available surfaces in all orientations, but for simplicity, the resource and cost assessments in this module only included south facing surfaces. Table 2 below summarises the current best estimates for the Accessible Resource for systems likely to produce electricity under 10p/kWh. These estimates are based on developed building integrated systems. The assumptions behind these figures are explained in sections E and F.

Discount Rate	England & Wales	Scotland	Northern Ireland	UK
8%	76	6.5	1.5	84
15%	4.0	0.2	0.0	4.2

Table 2: Accessible Resource in the UK (<10p/kWh future long term resource).

D. Environmental Aspects

PV offers the advantages common to most renewable technologies; it is emission-free and presents no hazards in operation. It avoids the noise problems associated with some other technologies and in the form of building integrated systems causes no land use or visual intrusion problems. In operation, PV systems require no fuel or cooling water and therefore raise no issues of supply or use of these resources.

The manufacture of PV materials involves the use of a number of hazardous materials, for example amorphous silicon requires the use of diborane, silane and silicon tetrafluoride gases. Recently use is also being made of cadmium, tellurium, copper and indium. In general, techniques exist to cope with the use of these materials, such as the use of multistage scrubbers. None of the environmental impacts of PV technology (materials, manufacturing processes, disposal etc.) appear prohibitive or beyond solution.

Health

Crystalline silicon PV modules pose no significant health hazards in their manufacture, operation or disposal. However thin film PV materials do use noxious or poisonous materials during manufacture, some of these can be present in the final products. The dispersal of these into the built environment may pose a health risk, even though the quantities embodied in individual systems are very small. The operation of small PV generation sets will carry the same risks to life as any other electrical installation but in the PV case DC voltages will be used and fewer people are familiar with DC technology.

The disposal of defunct plant has not been addressed yet on any significant scale. There should be few problems associated with silicon based systems since the materials are not hazardous but some of the newer devices use toxic materials such as heavy metals (e.g. cadmium). However the quantities are very small and disposal processes being developed for other markets (e.g. for batteries) should be directly applicable. It is not envisaged that decommissioning of plant will be a significant problem but further R&D will be required to establish methods and procedures if the PV industry grows to a significant size.

Environmental burden	Specification
Gaseous emissions:	None
Carbon dioxide	
Methane	
Nitrous oxide	
Chlorofluorocarbons	
Sulphur dioxide	
Nitrogen oxides	
Carbon monoxide	
Volatile organic compounds	
Ionising radiation	None
Wastes:	
Particulates	None
Heavy metals	<20 mg/GJ/year (cadmium etc. on disposal)
Solid waste	~ kg/GJ/year (glass etc. on disposal)
Liquid waste	None
Thermal emissions	None
Amenity/Comfort:	
Noise	None
Visual intrusion	Significant
Use of wilderness area	~ 5 sq.m/GJ/year for large plants
Other	None

Table 3: Environmental burdens associated with photovoltaics.

Since PV systems generate electrical power directly they can reduce the overall emissions from the power generation sector. The domestic-scale system outlined in section A would, on average, produce the emission savings shown in table 4 assuming that each kWh generated by photovoltaics displaces a kWh generated by the average plant mix in the UK generation system (for the year 1990) and thereby saves its associated emissions. The displaced emissions for a 50 kWp scheme on a non-domestic building are also shown.

Emission	Displaced Emissions (grammes of oxide per kWh)	Annual savings for example schemes (tonnes of oxide)	
		3 kWp (domestic)	50 kWp (non-domestic)
Carbon dioxide	734	2.2	37
Sulphur dioxide	10	0.03	0.5
Nitrogen oxides	3.4	0.01	0.17

Table 4: Emissions savings relative to average UK plant mix in 1990.

E. Economics

The costs of PV technology have been assessed in three different markets:

- remote power systems;

- grid-connected systems – central power plant;

- grid-connected systems – distributed, building integrated systems.

For each application the economic prospects have been assessed under a number of different cost scenarios: present costs, illustrative achievable costs, and best future costs.

Remote power systems
The economic assessment of PV for remote power systems was based on 10 example applications with loads varying in both size and pattern of demand. The results for these indicate a general pattern:

- for low loads (< 0.1 kWh/day) such as remote telemetry stations, holiday homes etc., PV appears already economically attractive;

- for loads in the range 0.1 kWh/day to 10 kWh/day, such as domestic lighting, refrigeration etc., PV may be an economic option, particularly for the smaller loads and those in areas of high solar resource;

- for high loads (> 10 kWh/day) the economics of PV appear generally unattractive when compared with diesel generation.

When market projections are taken into account the total contribution in terms of UK electricity supply is limited. Because of this, resource-cost data have not been prepared for this application.

Central power plants
Even under the best future cost scenario and the most optimistic assumptions on system efficiency and resource availability, PV for central power generation still appears to be an uneconomic option for the UK. Costs for sites in Southern Europe appear to be more promising. The Accessible Resource however, even in the UK, is quite large but the cost per unit of delivered electricity is estimated to be greater than 10p/kWh even on the best projections.

Building integrated systems
Building integrated systems offer potential advantages in terms of reduced costs for land and support structures and the avoidance of transmission losses. If all of these are taken into account, building integrated systems appear attractive under the best future cost scenario. More detailed studies of daily and seasonal variations in PV performance, electricity demand patterns, tariff structures and the scope for substituting conventional

building materials with PV panels are required to confirm this assessment.

Data for typical current and projected future systems are given in table 5 for both central generation and building integrated systems.

Plant type: building mounted systems 1 to 500 kWp central systems > 500 kWp			
	Range for available plant	Range for example plant (2010)	(2025)
Capital cost (£k/kW):			
Central plants modules	1.7 to 3	0.8 to 1.0	0.3
Balance of system (BOS)	2.2	1.0	1.0
Retrofit buildings modules	1.7 to 3	0.8 to 1.0	0.3
Balance of system (BOS)	0.7	0.4	0.4
New buildings total system allowing for avoided costs	1.5 to 1.7	0.8 to 0.9	0.4 to 0.5
Annual recurring costs (£k/kW)	Not known, assumed nil	Not known, assumed nil	Not known, assumed nil
Performance data:			
Plant availability	95%		
Average load factor	12 to 15%		
Type of operation	Intermittent and unpredictable		
Onsite power losses	20%		
Other:			
Construction time			
Central plants	1 to 2 years		
Building mounted	0.2 to 1.0 years		

Table 5: Cost and performance data.

The ranges in module cost cover the various material options from thin film to crystalline silicon. The economic analyses in section F assume the lowest cost for each case.

F.   Resource-Cost Curves

Resource-cost curves have been prepared for both central generating plant and building integrated systems. At the present time there is little data on which to base these calculations and the crude approximations and assumptions used are summarised below. Since PV technology is expected to undergo rapid development, three sets of data have been prepared for each application. The first uses performance data for current technology systems, the second uses that for a projected system performance which might be achieved within one to two decades, the third assumes long term "ultimate" systems. In addition, for simplicity, the UK has been divided into two solar resource regimes – a best site solar resource (3.0 kWh/sq.m/day) and an average site solar resource (2.5 kWh/sq.m/day). This division is shown in table 6.

Region (land area and buildings)	% Best site	% Average site
England & Wales	50%	50%
Scotland	0%	100%
Northern Ireland	0%	100%

Table 6: Best site and average site solar resource.

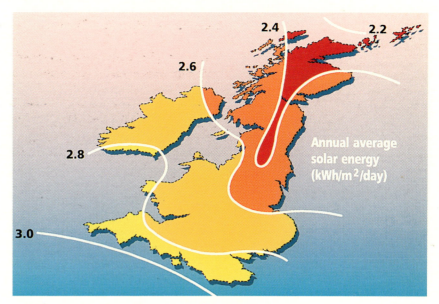

Figure 2: Resource map of UK solar energy.

The output per installed Wp for the best site would be 1 kWh/Wp/year assuming a load factor of 15%, system availability of 95% and system losses of between 1 and 20%. For the average site the output per installed Wp would be 0.8 kWh/Wp/year.

Performance in terms of Wp/sq.m:

Current technology	50 W/sq.m
Future achievable Technology	100 W/sq.m
Best future technology	150 W/sq.m

(Note that these relate to module efficiencies of 5, 10 & 15% respectively. This corresponds to the performance of current amorphous silicon and future thin film technologies. Crystalline silicon is more efficient, up to 16% currently, but at a higher cost. In the long term it is believed that thin film modules will be the cheapest technology and best suited to applications such as building integrated systems.)

While the performance of PV systems (and therefore the unit generation cost) is directly related to the amount of solar radiation, it is likely that actual costs will be dominated by a number of other issues which are difficult to take into account in this simple analysis.

		1990	2005	2025	(2030)
Assuming current trends continue	Wp/sq.m	50	88	138	(150)
	£/Wp	4	2.5	1.25	(1)
Assuming more environmentally sensitive policies and practices are adopted	Wp/sq.m	50	112	150	(150)
	£/Wp	4	1.75	1	(1)

Variation in system cost and performance with time.

The development of PV technology is driven by the R,D&D activities of many national governments and private companies. In order to develop resource-cost curves for the two dates of 2005 and 2025, assumptions need to be made on the rate of improvement of PV efficiencies and costs. Future environmental policies and practices will have a significant impact on the effort put into developing PV technology and its markets. Two views of the future have therefore been considered. The first assumes that current trends continue; the second that there is a move towards more environmentally sensitive policies and practices. The development of PV technology and its market is assumed to respond accordingly.

The main difference between these two views of the future is the rate of progress between 1990 and 2005. The ultimate long term performance (2030) is assumed to be the same in both cases. The underlying premise for this is that moves towards more environmentally sensitive practices and policies (higher fuel prices and stringent emissions constraints) will accelerate the rate of development but that the fundamental long term achievable performance will not change.

Central generating plant
The Accessible Resource based on land area has been estimated as:

England & Wales	76,000 sq.km
Scotland	45,000 sq.km
Northern Ireland	9,200 sq.km

These estimates make allowance for built-up areas, roads, lakes & rivers, National Parks & SSSIs etc. and give an upper limit to the feasible resource. In order to estimate the Accessible Resource, they have been further reduced by a factor of 1,000 to allow for unfavourable gradients, packing densities, access and other infrastructure requirements. This reduction factor is largely arbitrary and is significantly larger than that used in the wind energy module, for example. This reflects the nature of large scale PV which precludes any other use of the land taken up by the installation. Even with this large reduction in land availability the Accessible Resource for the UK is still significant at around 15 TWh/year.

The resource has been calculated as:

Annual Resource = Land area(/1,000) x Wh/Wp/year x Wp/sq.m

(The cost data has been calculated from the information on system performance in section E).

Building integrated systems
There are two components to the Accessible Resource for systems sited on buildings – retrofitting to existing buildings and integration into new buildings. These need to be considered separately as the costs are likely to be different. For retrofit systems, only savings in the PV support elements can be included whereas for new-build systems it is expected that some savings in the building material costs can also be made.

The outputs per house and per square metre of non-domestic roof space are as follows.

Houses (kWh/year):

Current technology	Achievable technology	Best technology
3,300	6,600	9,900

(Based on house plan area of 40 sq.m, modules inclined at 50°.)

Non-domestic (kWh/sq.m/year):

Current technology	Achievable technology	Best technology
50	100	150

(Note that cladding of walls may be as attractive an application as roof systems, particularly for non-domestic buildings. However, it is not easy to identify the available wall areas since orientation and shading will have a significant impact. For simplicity this analysis only includes roof areas; consequently it may be conservative.)

The resource is determined by the number of houses, area of roofing and the geographical distribution. The estimated Accessible Resource is as follows.

Retrofit resource:

	England & Wales	Scotland	Northern Ireland
Number of houses	20 M	2.1 M	55 M
Area of non-domestic roofing (sq.m)	240 M	30 M	7 M

The resource is determined by:
Annual resource = (No of houses x output/house (Wh/year) x $\frac{1}{4}$) + (Area of non-domestic roof x output of roof (Wh/sq.m/year) x $\frac{1}{2}$).
(The factor of $\frac{1}{4}$ for houses and $\frac{1}{2}$ for non-domestic buildings is to allow for the likely fraction facing south).

Newbuild houses and commercial buildings need to be considered separately since the displaced building costs will be different.

New-build houses:

	England & Wales	Scotland	Northern Ireland
Number of houses/year	136,000	10,000	4,000

The resource in a given year is determined by:
Annual resource = No houses/year x output/house x $\frac{1}{4}$.
(The factor of $\frac{1}{4}$ is again to allow for orientation).

New-build non-domestic:

	England & Wales	Scotland	Northern Ireland
Area of roofing (sq.m)/year	2.2 M	300,000	150,000

The resource in a given year is determined by:
Annual resource = Area of roof(sq.m)/year x output/sq.m x $\frac{1}{2}$.
(The factor of $\frac{1}{2}$ is again for orientation).

The Maximum Practicable Resource has been determined from a subjective assessment of the number of systems that could reasonably be installed by 2005 and 2025 respectively. This assessment has taken account of such factors as manufacturing capacity, demonstration activities, availability of installation services, etc. For applications which achieve low costs (< 10 p/kWh), the Maximum Practicable Resource in 2025 tends towards the Accessible Resource.

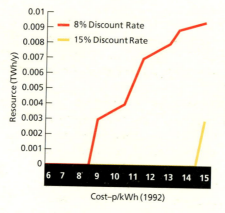

Figure 3: Maximum Practicable Resource-cost curves for photovoltaics in 2005 assuming current trends continue.

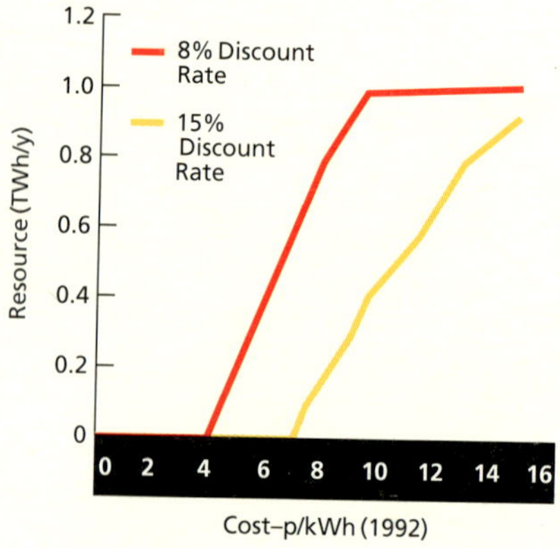

Figure 4: Maximum Practicable Resource-cost curves for photovoltaics in 2025 assuming current trends continue.

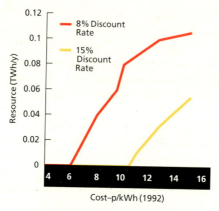

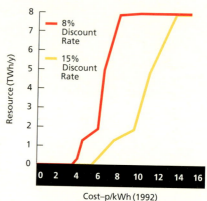

Figure 5: Maximum Practicable Resource-cost curves for photovoltaics in 2005 assuming more environmentally sensitive policies and practices are adopted.

Figure 6: Maximum Practicable Resource-cost curves for photovoltaics in 2025 assuming more environmentally sensitive policies and practices are adopted.

The data assuming current trends continue was used for the modelling of the Composite, Low Oil Price, High Oil Price and Shifting Sands scenarios and that assuming more environmentally sensitive policies and practices was used for the modelling of the Heightened Environmental Concern scenarios. The results of the modelling exercise are presented in Section H.

G. Constraints and Opportunities

Technological / System

The development of PV devices and systems has undergone continuous improvement over the last decade. This has resulted in device efficiency improvements and reductions in the unit cost, or £/Wp. Most of this improvement in performance and cost is a result of developments in the device architecture and manufacturing techniques for silicon-based devices. As mentioned in section A, there are new materials currently under investigation or at the pilot production stage which, if their promise is borne out, will enable a further significant reduction in the £/Wp of PV cells to be achieved. Attainable targets for £/Wp of the order of 1.0 are thought be reasonably certain over the next decade, with values as low as 0.3 considered possible in the longer term, (module costs).

It is likely that PV technology will continue to enjoy spin-off benefits from developments in the microelectronics industries generally. The current world-wide investment in new materials and devices by these industries is very large and technological breakthroughs applicable to PV systems are likely.

If these performance targets are realised then PV systems will become attractive options for small-scale distributed systems integrated into buildings. There are, however, other technical issues which may constrain the level of deployment. These are addressed in outline below.

The cost of the PV module is only a part of the overall cost of a system. The remainder or Balance of System (BOS) elements are currently about 50% of the total cost, depending on the size of system under consideration. As developments in the module components occur a greater fraction of the cost will be borne by the BOS elements. The main components of the BOS are the structural supports, the inverter and voltage regulation equipment, interconnections, isolation gear, etc. The use of PV systems in buildings should reduce the structural requirements but the other components are unlikely to undergo major developments which could result in cost reductions. There may therefore be a plateau in energy costs below which PV systems are unable to go. This is probably one of the major factors which will prevent large-scale central generation becoming economically attractive in the UK.

The economics of small-scale PV generation in buildings will be judged by the individual installer/consumer. The attractiveness of PV systems is likely to be largely determined by factors such as the load profile of the consumer - in other words, can he use all the power 'in house', or would most of it be exported to the grid? Buildings with significant daytime loads such as air-conditioning, core lighting, etc. would probably be able to make use of all the electricity PV systems could generate. This would displace peak rate electricity and would therefore currently be worth about 7p/kWh to the consumer. Any special technical requirements for connection to the grid and the addition of two-way metering would increase the overall cost of the system.

Some technical developments in building electrical systems may be beneficial to PV power installations. For example, low voltage, direct current lighting systems are under development. These could form ideal loads for PV systems and remove the need for inverters and improve the economics overall.

There is little information as yet on the long term reliability and maintenance of PV systems, particularly in urban environments. Current experience indicates that system availability should be quite high (~ 95%) and that maintenance needs are modest. There is more data on system efficiencies. Values as high as 85% have been measured in demonstration systems but 75% to 80% is a reasonable estimate. Significant advances in performance are unlikely but components specifically designed for PV applications and well matched to system capacity may well produce some improvement.

The major opportunity for PV is the expansion of the existing markets as more end use applications are developed. As explained in section B, the largest current markets are for consumer goods and remote power.

UK market growth potential
The consumer product market already exists; rapid growth of 25% every 5 years has been assumed up to 2010, with slower growth thereafter. By 2025 the annual sales are expected to be just below 1 Wp/household.

An assessment has been made of the likely market growth for PV remote power to the year 2025 for a range of system/application sizes. The results are summarised below:

- for applications of less than 2 Wp (e.g. telemetry), the maximum was estimated as 10,000 installations/year of average power 1 Wp;

- for applications of 2 to10 Wp (e.g. building lighting, small telecommunications, larger telemetry and navigational aids); a small but growing market estimated at 1,500 systems/year in 1990, growing to 40,000/year by 2025;

- for systems of greater than 10 Wp (e.g. remote building power), a small market with an estimated 50 systems of average power of 1,000 Wp and 1,000 systems of average power of 80 Wp in 2015, with steady growth to 2025.

The total forecast size of the market is shown in table 7 below:

	1990	2000	2015	2025
< 2 Wp	1	3	10	10
2-10 Wp	9	22	85	230
> 10 Wp	10	25	125	360
Total	20	50	220	600

Table 7: UK remote power applications by PV power requirements. (Annual sales in kWp).

The development of the grid-connected market is much more difficult to forecast as there is no current market. It has been assumed that no large systems (greater than 100 kWp) are installed before 2015 but a small number start appearing after that. Smaller systems will develop a small market (200 kW/year) by 2000, with steady growth thereafter. The proposed installed capacity would correspond to PV systems being installed on 1 to 2% of new buildings over the period 1990 - 2025.

Current market projections are subject to a wide range of uncertainty, but the broad conclusions are that PV is likely to make a modest contribution to UK electricity supply by 2025.

World PV markets are expected to grow very substantially as more niche applications become economically attractive, for example remote power in developing countries and regions with particularly favourable solar conditions. Many estimates have been made of the future size of the global PV market; a figure of 1.8 GWp/year by 2025 has been suggested (the world market in 1992 was 60 MWp). Clearly this represents a significant export potential for the UK PV industry. The current geographical distribution of sales is: Europe 23%; Asia and Pacific 22%; N America 21%; Africa 13%; Japan 10%; S America & Caribbean 7% and the rest of the World 4%. This breakdown may change in the future if grid connected systems take over from remote power as the largest market. In the short term, however, the remote power market in developing countries is expected to show the fastest growth. It is estimated that in 1992 the UK PV industry supplied about 6 MWp (10%) of the world market for PV goods. A large fraction of this, however, used components sourced from outside the UK.

Institutional

PV systems are currently expensive for most applications where a grid supply is available. As costs of systems fall over the next decade or so, an increasing number of applications may become cost-effective. It will be necessary to counter the view that PV systems are always expensive and therefore can have no role in the UK. This could be tackled by building demonstration schemes, but it will be necessary to have reliable components and ideally "off the shelf" systems available to prove the case. This will require investment by the PV industry in the right components and will probably require partnerships between the PV component manufacturers and the relevant building component manufacturers and suppliers.

It will be important to ensure that components developed for building PV systems fully comply with the requirements of safety and other standards. This area will be covered in future by the European Construction Products Directive and the components developed will need to comply with this. The conservative attitude with which the construction industry usually views new technology may well prove to be a significant barrier.

The connection to the grid system of large numbers of small independent generators could have significant implications for the whole electricity industry. For example it will be necessary to ensure adequate safety standards in terms of both system and component quality and provide isolation mechanisms in case of system or grid failure. A suitable means of metering small amounts of exported power would need to be devised and the costs and tariffs associated with this worked out. It may be necessary to provide specific mechanisms to enable such small-scale generation schemes to be introduced. Most building systems would be below 50 kWp installed capacity and so would be too small to be considered under support schemes such as the present NFFO.

There is no information at present in the UK on the public perception and acceptability of PV systems on buildings. A German initiative to install systems on domestic houses (see section II B) has proved to be extremely popular but it does carry a 70% subsidy.

H. Prospects – Modelling Results

The results of the MARKAL modelling exercise for PV using the resource-cost curves described in section F are summarised below in tables 8, 9 and 10.

Scenario	Discount rate	Contribution, TWh/year			
		2000	2005	2010	2025
HOP	8%	0	0	0	0.54
	15%	0	0	0	0
	Survey (10%)	0	0	0	0.30
CSS	8%	0	0	0	0.54
	15%	0	0	0	0.30
	Survey (10%)	0	0	0	0.30
LOP	8%	0	0	0	0.30
	15%	0	0	0	0
	Survey (10%)	0	0	0	0
HECa	8%	0.005	0.03	0.09	7.2
	15%	0	0	0	1.8
	Survey (10%)	0	0	0.09	4.8
HECb	8%	0.005	0.03	0.09	7.2
	15%	0	0	0	1.8
	Survey (10%)	0	0	0.09	4.8

Table 8: PV decentralised potential contribution under all scenarios.

Table 8 shows the predicted uptakes of PV generated electricity (in TWh/year) from 2000 through to 2025, over a range of future energy scenarios and for three discount rates. A number of clear trends emerge which can be summarised as follows.

- For the CSS, LOP & HOP scenarios, decentralised PV makes no contribution until 2025. The predicted contribution only exceeds 0.5 TWh/year under the application of an 8% discount rate in the CSS and HOP scenarios. This represents about 50% of the UK Maximum Practicable Resource set out in section C. In the LOP scenario, the PV contribution is reduced to about 1/3 TWh/year.

- Centralised PV stations never make any contribution under any scenario.

- For the HECa and HECb scenarios, decentralised PV starts to make a small contribution as early as 2000 under an 8% discount rate and this grows to over 7 TWh/year by 2025. This contribution represents about 90% of the UK Maximum Practicable Resource of 8 TWh/year estimated in section C. Under the higher discount rates there is no contribution until 2010 but there is still a significant contribution by 2025 of nearly 2 TWh/year (15% discount rate) and nearly 5 TWh/year (survey discount rate).

These results are based on data which assumes a steadily decreasing cost per kWh for PV systems all the way through the time period considered. The significantly increased contributions predicted for the HEC scenarios result from the higher marginal cost of electricity used by the model in those scenarios (as a result of emissions restrictions etc.) and the faster rate of development of PV systems assumed in the resource cost curves used for this scenario. To test the sensitivity of the results, the more pessimistic resource cost curve used in the other (CSS, LOP & HOP) scenarios was also applied in the HEC cases. With that change of input data the early PV contributions under the HEC scenario at 8% discount rate disappear and the 2025 contribution falls to about 1 TWh. It might be concluded therefore, that a robust estimate of the PV contribution under any scenario is in the range of 0.5 to 2 TWh/year by 2025.

Environmental burden			Incremental contribution (TWh/year)			
	Constraint	Base case scenario	2000	2005	2010	2025
Carbon dioxide	- 5%	CSS	0	0	0	0.08
	- 5%	HOP	0	0	0	0.24
Carbon dioxide	-10%	CSS	0	0	0	0.41
	-10%	HOP	0	0	0	0.24
Sulphur dioxide	-40%	CSS	0	0	0	0
	-40%	HOP	0	0	0	0
Nitrogen oxides	-40%	CSS	0	0	0	0.44
	-40%	HOP	0	0	0	0.24

Table 9: PV decentralised incremental contribution under environment constraints. (TWh/year above base case scenario contribution).

Table 9 shows the effect of the sensitivity studies which were carried out to assess the effect of variations in the environmental constraints applied to the CSS & HOP scenarios. A 10% decrease in the allowed carbon dioxide emissions per decade enhances the PV contribution by some 240 GWh/year in 2025 for the HOP scenario but under the CSS scenario the increase is much greater at 411 GWh/year. Similar effects are seen under the nitrogen oxides constraint but the sulphur dioxide constraint has no effect.

Incremental contribution (TWh/year)				
Discount rate	2000	2005	2010	2025
8%	0	0	0	0
15%	0	0	0	-0.2
Survey	0	0	0	0.08

Table 10: PV decentralised incremental contribution under Shifting Sands assumptions (TWh/year above CSS contribution).

Table 10 shows the effect of the "Shifting Sands" change to the CSS scenario. This has little effect on the PV contributions because the hikes in fuel prices occur before PV has become economically viable under these conditions.

II. Programme Review

A. United Kingdom

The PV community in the UK is relatively small and consists of the following:

- module manufacturers;

- PV component suppliers (mainly consumer goods, pumps and telecommunications equipment);

- related industrial PV research;

- BOS component manufacturers and suppliers (batteries, standard electrical devices);

- University PV material R&D laboratories;

- testing facilities;

- consultants (mainly involved in overseas markets).

The UK academic groups working in the PV field have an internationally acclaimed track record of achievements in the development of new materials and devices. There are major groups working at the University of Northumbria at Newcastle, Imperial College London and Southampton University along with several other smaller groups. The main funding support for this work comes from the Science and Engineering Research Council, the Link programme (DTI/SERC) and the European Commission.

BP Solar are among the world market leaders in manufacturing crystalline and poly-crystalline PV modules and PV appliances such as remote power systems. However they do not currently have a UK manufacturing base. BP are also engaged in research into new thin film based modules and high efficiency crystalline modules. SOLAPAK manufacture amorphous silicon modules and produce a range of consumer and commercial products.

Within these groups however, there is little direct experience of installing and operating grid connected PV systems within the UK. The principal exception being the European Union/DTI sponsored demonstration plant at Marchwood, installed and operated by BP Solar and the (then) CEGB. This installation has now been dismantled. The only examples of building integrated systems in the UK at present are the SOLAPAK houses at the Milton Keynes Energy Park equipped with PV modules in the conservatory roofs, as shown in figure 7.

The main priorities of the PV module manufacturing industry are the development and introduction of new manufacturing techniques to enable continuous production of higher efficiency cells at lower unit cost. This is mainly centred around the thin film materials, although work is also being carried out to optimise the performance of crystalline modules. The balance of supply and demand is constantly changing within the world PV market as new manufacturing plant is installed and demand in the various market sectors fluctuates. These fluctuations can significantly affect the retail value of basic modules on the open market. If there is a period of oversupply then the emphasis can shift to systems engineering which will result in added value rather than high volume module manufacturing. This situation may change if the size of the world market continues to grow.

The development of a larger market is also the priority of consulting organisations in the industry, in particular expansion of the remote power systems market in developing countries. The need here is for demonstrations of reliable systems which meet the needs of the local populace. The organisations funding technology placement in developing countries, such as the World Bank, World Health Organisation, UNDP and the various national aid agencies, need evidence that PV can offer practical cost-effective solutions.

Figure 7: House incorporating photovoltaic cells, Milton Keynes.

The Government's Programme

Photovoltaic systems were not included in Energy Paper 55. The recent review of the technology showed that significant developments have occurred in terms of both system performance and unit manufacturing costs. As a result of this a programme on PV systems has been instigated within the DTI programme. The objectives of this initial programme on PV can be summarised as follows:

* to assess the potential for distributed PV electricity generation in the UK;

* to identify the barriers to the installation and use of PV systems.

Without Government involvement, the development of PV systems for building integration is likely to be very slow. The main reasons for this are the conservatism of the construction industry with regard to new technologies and the perceived risks of applying new components and techniques. New products will need to be developed for this market which will require investment by the PV and building industry. In addition certification and testing procedures will need to be established. There is a need therefore, to determine what the technical and market barriers facing this application are likely to be. If the energy, environmental and wealth creation prospects for PV technology can be shown to be favourable then the relevant industries might be encouraged to make the required investments.

The main elements of the DTI programme can be identified as:

- an assessment of the potential for building integrated PV systems in terms of suitability of existing and future structures, likely generation output etc;

- the determination of the main design issues for PV components and systems for building-integrated applications;

- the identification of the design and construction issues which architects and services engineers will need to face, including the requirements of building regulations, electrical safety standards, etc;

- an analysis of the BOS components likely to be required as part of a building integrated PV system and an assessment of weak links in the system and the possibilities for improvement;

- an analysis of the demand load profiles in a range of building types and an assessment of the ability of PV sourced power to satisfy these demands;

- an assessment of the utility interface issues which may constrain the widespread application of distributed grid connected systems;

- an analysis of the likely cost of building PV components in the near to mid-term, including the impact of cell cost reductions, re-design issues etc.

Central to each area of activity is the identification of barriers to the widespread uptake of the technology and the steps needed to overcome these. In this work the assumptions used must be made clear so that the robustness of the conclusions reached can be assessed by the target audience.

The SERC supports work on semiconductor development including basic PV device R&D, through its own programme and also in conjunction with DTI through the LINK programme. In addition, work on PV systems for electricity generation are included in the activities of the International Energy Agency and the European Union. Under the CEC DGXII programme, several countries have operated pilot plants for a number of years with capacities in the 20 kWp to 200 kWp range (including the Marchwood plant referred to in section II A). In addition, the JOULE programme of non-nuclear energy research is supporting a significant activity in new PV materials. The Demonstration and THERMIE programmes of DGXVII have also supported a number of PV technology projects covering a range of applications from small remote power through to large scale central power plants.

The International Energy Agency has initiated a new Implementing Agreement to co-ordinate R,D&D effort on PV power systems. Work on the architectural integration of PV systems into buildings is also being carried out within the IEA Solar Heating & Cooling Programme.

B. Other Countries

There are significant PV programmes underway, or about to commence, in several countries.

The countries with the highest profile in PV research and application tend to be those with significant indigenous manufacturing interests in PV. These are USA, Germany, Japan, Italy, France, Spain, and Australia. Other European countries with interests in PV include Switzerland, Austria, Holland, Finland and Ireland.

Worthy of particular mention is the German '1,000 Roofs' programme to install small (~1 to 3 kWp) grid connected PV systems on a large number of private houses. The original target of 1,000 installations has been increased due to the incorporation of the former East German regions and also the favourable response from the public. The installations are being subsidised by both the Federal and regional governments to a limit of 70% of the cost.

There are many examples of remote site applications from Alpine mountain huts to hotels, restaurants and isolated island communities. Many of these are hybrid systems with diesel backup.

There are several ongoing programmes to investigate the potential of large scale central PV plants. One of these is the PVUSA (PV for Utility Scale Applications) programme in the USA. Another is a 3.3 MWp installation in Italy.

Photoconversion

I. Technology Review

A. Technology Status

Photoconversion is a generic term describing processes which capture sunlight and convert it into either electrical power, heat or a chemical fuel. The processes used can be biological, chemical or electrochemical. This appraisal covers photobiology, photochemistry and photoelectrochemistry but excludes solid semiconductor photovoltaics which are addressed in a separate module. The concept is interesting as a potential energy resource because:

- it relies on the sun's energy which is freely available to drive the reactions;

- a substrate for many reactions is water which is widely available;

- it has low ecological impact;

- it does not add to atmospheric carbon dioxide levels.

At present photoconversion processes which convert sunlight directly into usable electric power or chemical fuels are largely confined to the research laboratory. From this research base the requirements for successful operation of industrial scale plant have been predicted, making broad assumptions and educated guesses where necessary. Consequently a broad band of uncertainty exists. The basic reaction principle is relatively simple. When sunlight is absorbed by a photoconverter a transient 'excited' or energy rich storage state is produced; this captured energy is then converted into some usable form, such as electricity.

Photobiology is the study of how light interacts with biological systems. In photosynthesis the captured solar energy is used to drive a series of chemical reactions, the end product of which is sugar. Photosynthesis is carried out by plants, algae, cyanobacteria and photosynthetic bacteria. Natural photosynthetic systems such as these have a number of advantages over synthetic systems (i.e. chemical or electrochemical) including:

- self replication;

- the ability to survive under adverse conditions;

- longer lifetimes.

Photochemistry is the process whereby solar energy is absorbed by dyes, pigments or semiconductors and then used to drive a photochemical reaction. In some cases the energy could be used to produce fuels and chemicals, hence storing the energy for future use. In other circumstances the energy may be released as heat. Most photochemical reactions are multi-step with only the first step driven by light energy and the subsequent steps occurring spontaneously in the dark. The advantage of multi-step reactions is that the final products can be produced in separate locations so reducing the chance of back reactions which can decrease the overall yield.

Photoelectrochemistry is the science of the action of light on electrodes in batteries (electrochemical cells) – more specifically electron transfer reactions occurring at illuminated electrodes. Two electrodes are involved –

a positive and a negative – and separate reactions occur at each, splitting a substrate, such as water. The two electrodes provide separate locations keeping the breakdown products apart. The end products of the photochemical and photoelectrochemical conversion systems described in this module are either electric power or hydrogen. Such systems can also fix carbon dioxide from the atmosphere.

Following a comprehensive review of this area in terms of the technical status and engineering prospects, two systems – one electrochemical and one biological – have been identified as sufficiently developed at this stage to warrant further consideration. The first system is electric power generation by electrochemical photovoltaic systems; the second the photobiological production of hydrogen. Photoelectrolysis is less well developed than these two systems, but more advanced than the following systems which require considerable further research before their commercial prospects are worth assessing:

- photoelectrochemical (PEC) storage cells (excluding ECPV cells);

- photogalvanic (PG) cells;

- photo-assisted electrolysis (PAE) cells;

- semiconductor suspension systems;

- photochemical electron transfer with micro heterogeneous catalysts;

- hybrid systems.

Electrochemical Photovoltaic Systems
PEC cells are similar to ordinary batteries in composition, consisting of an anode or oxidising reaction electrode, a cathode or reducing reaction electrode and an electrolyte solution. In normal batteries the reactions are spontaneous, in PEC cells they are driven by sunlight absorbed by either the electrode(s) or the electrolyte.

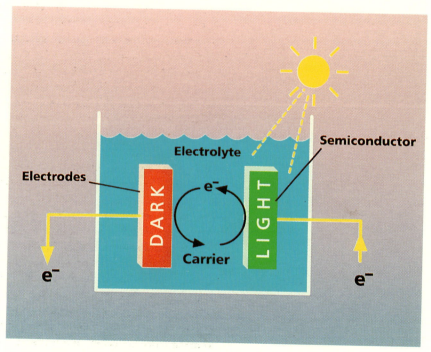

Figure 1: Schematic of an electrochemical photovoltaic cell.

Electrochemical photovoltaic cells (ECPV) are a type of PEC cell. They are functional equivalents of conventional solid state semiconductor photovoltaic (PV) cells which produce DC electric current on illumination. Although much less highly developed than solid state photovoltaic cells, the efficiency of ECPV cells is comparable and they may have potential for reduced production costs. The main difference between the two systems is that the light reactive centre of photovoltaic cells is embedded within the semiconductor whereas in ECPV cells it is on the surface. This means that a rough electrode surface, and hence crude and inexpensive design, is possible for ECPV, whereas with a photovoltaic device effectiveness depends upon a uniform and accurate junction inside the semiconductor crystal matrix. Also in the ECPV case the junction can be illuminated directly but a PV cell requires a light absorbing top layer. A further advantage offered by ECPV is that semiconductors with only one of the two possible conductivity options can be utilised. PV cells require both options. They should also, for example, be able to tolerate electrode contact with loss of function in only a small localised area. All of these factors are significant in terms of cost reduction because they mean that manufacturing requirements could be simplified. Several ECPV cells have been constructed with efficiencies above 10%, rivalling the performance of single junction photovoltaic cells. However ECPV cells have only been tested in small area configurations so the extent of power loss on scale-up is uncertain. Laboratory-based systems have been developed with a 10% conversion efficiency.

On the negative side, most semiconductor electrodes are vulnerable to photo-corrosion (although surface coatings can provide protection in some instances) and electrolyte solutions are degraded by atmospheric oxygen. As cells have only been operated to date for short periods of time (a maximum of several months) durability is not established.

Photobiological Production of Hydrogen
To date a large number of workers throughout the world have developed laboratory-scale photoheterotrophic bacterial systems capable of continuously producing hydrogen, e.g. continuous hydrogen production by free living cell cultures for up to 80 days, based on a lactate containing waste. Hydrogen generation by photoconversion is obviously light sensitive, but research has indicated that lower light intensities produce a comparable hydrogen generation rate per gramme of biomass. This implies that the deeper, more shaded areas of the reactor may be as productive as the surface. The efficiency of the conversion of solar energy into hydrogen by the anaerobic photosynthetic bacteria has been estimated to be 5% to 6% and with optimisation the National Renewable Energy Laboratory in America believes that efficiencies of 10% to 15% could be achieved.

An industrial-scale system might be based on plug flow semi-continuous operation where the substrate enters horizontally at one end of the vessel and is then drawn off at the opposite end. The substrate concentration would decrease along the length of the reactor vessel as bacterial growth and metabolism occurs. The withdrawn material would undergo a separation stage where biomass (bacteria) would be separated from the effluent and recycled to the front of the reactor. This would enable high bacterial activity to be maintained. As the conversion of the organic substrate would be incomplete some form of residual effluent treatment would be required, incurring some environmental burden. Also, although oxygen would not be an end product, carbon dioxide would be, so a gas scrubbing system would be necessary. This may not necessarily require venting since collection and utilisation would be possible. It should be noted that carbon dioxide produced by such a system would have been taken up

Current Status	Research
UK installed capacity	None
World installed capacity	Not known
UK industry	Several university research groups

Table 1: Summary of photoconversion development status.

and used by the organism for its own growth. Consequently there would be no net contribution to external carbon dioxide levels. A possible system for photobiological hydrogen production might be located at a municipal waste water treatment site where it could be interfaced with an existing anaerobic digestion stream. This would facilitate access to a suitable substrate and provide both an effluent disposal route to the anaerobic digestion system and a source of heating. A similar approach could be taken with systems based on other waste streams.

Hydrogen is an environmentally attractive fuel as its combustion produces no carbon dioxide, just water and NO. It contains more than twice the energy content per unit weight of other hydrocarbon fuels and it can be used directly or be added to natural gas at 10% (by volume) with no requirement for retrofitting. In terms of drawbacks, hydrogen is a highly flammable gas. Hydrogen flames are almost invisible although extremely hot; also hydrogen can readily diffuse through very small apertures in the containment or distribution system

B. Market Status

At the present time none of the photoconversion processes described above have reached the stage of commercial viability in the UK. Commercial scale-ups of variants of some of these processes have been attempted in a few countries, the most notable being the growth of Dunaliella algae in saline tanks in Israel. The ECPV approach has been developed furthest in Switzerland and there are plans there for a limited pilot production facility.

C. Resource

The resource for photoconversion systems will be determined by:

- the solar energy resource;

- the system efficiencies;

- the means of deploying the systems.

At the present time there is insufficient information on which to base a detailed resource assessment. There do not, however, appear to be any overriding factors associated with the UK solar resource which would preclude operating the systems outlined – although it is generally true that a higher solar resource will enhance the system performance and reduce the costs of the energy produced. It is possible to envisage electrochemical photovoltaic cells being deployed in similar ways to conventional photovoltaic systems. The Accessible Resource in the long term (beyond 2030) is therefore likely to be of the range of 50 to 100 TWh/year.

Photobiological hydrogen systems are likely to be deployed in parallel with other waste treatment processes. The resource will therefore be determined by the economics of adding this technology to existing installations. This is likely to be in discrete systems at a limited number of geographical locations. The low light levels thought to be required for some of these processes suggest that the solar energy resource is of less consequence than many of the other factors, such as availability of suitable waste streams.

D. Environmental Aspects

A detailed analysis of the environmental factors associated with these photoconversion processes has not yet been carried out. An initial assessment of the possible environmental burdens is shown in table 2. It is not yet possible to determine the levels of the various environmental burdens as a function of the GWh/year of generated energy. None of the possible burdens is, however, unique to photoconversion processes and should not therefore represent an insurmountable barrier.

Environmental burden	Electrochemical PV cells	Biological H$_2$ production
Gaseous emissions:		
Carbon dioxide	None	Yes*
Nitrous oxide	None	Yes+
Hydrogen sulphide	None	Yes
Ionising radiation	None	None
Wastes:		
Heavy metals	Possible	No
Solid waste	Yes	No
Liquid waste	Yes	Yes
Thermal emissions	None	Yes
Amenity/Comfort	None	None

Table 2: Environmental burdens associated with photoconversion processes.
   * Previously removed from the environment for growth of organisms, therefore no net gain.
   + If hydrogen product is burnt.
   Note: This technology is still at an early stage of assessment, detailed figures are not yet available.

Health

Many of the possible choices for redox couples are toxic; the safety aspects of any application, therefore, would need to be borne in mind.

Effluent will be produced and this will require acceptable disposal routes. A secondary clean up procedure may be required; this will add to costs and may introduce an environmental burden.

E. Economics

This technology is in its infancy and there are currently no exemplar systems in the market. At the current level of knowledge only speculative cost estimates for the systems outlined here can be given. A system in the UK producing 10 MW$_e$ might have the following costs.

System	Efficiency %	Cost £/sq.m	Cost £million/MW
ECPV	0.15	40	2.5
Biological hydrogen	0.08	40	5

For systems with a 25 year lifetime these costs lead to a generation cost of around 20p/kWh, depending on the discount rate used. The actual costs will be dependant on the efficiencies of full scale systems, operating and maintenance costs and – for biological hydrogen production - any waste treatment credits or penalties which could be applied. None of these factors has yet been assessed in any detail.

F. Resource-Cost Curves

Resource-cost curves have not been prepared for this module as there is currently insufficient detailed information on which to base an analysis.

G. Constraints and Opportunities

Technological / system

There are a number of issues relating to the manufacture and operation of ECPV cells which require attention. The more important of these are summarised as follows.

- Electrolyte degradation is a major concern. Ideally reprocessing requirements should be avoided as these would increase the complexity of the system.

- Suitable semiconductor surfaces need to be prepared to facilitate electron transfer.

- Chemical diffusion between light and dark electrodes will impede electron flow limiting current density due to concentration effects. Consequently voltage will be needed to drive the electron transfer reactions and this will reduce the energy yield. This imposes a requirement for an extremely thin electrolyte layer if the process is to minimise the internal voltage requirement and hence maximise the output.

- Achieving good physical stability of the system is important. Plastic, a potential container of the system, often contains UV resistant materials. This would be unhelpful for light absorption systems. A further requirement would be oxygen exclusion because of corrosion problems which interfere with chemical stability. Rigorous oxygen exclusion suggests high containment costs or the use of resistant materials which would increase the electrode costs. Striking a balance between operational requirements and costs presents a challenge.

- An extended electrolyte life would remove a potential obstruction to commercial production. If a long enough electrolyte life could not be achieved the electrolyte would need to be withdrawn, reprocessed and regenerated. Removing the electrolyte could be difficult.

- ECPV cells have the potential to develop self-limiting short circuits so that, if the two electrodes should come into contact with one another, damage and hence loss of function would be limited to the small area affected. Overall functioning would therefore be maintained. This gives ECPV a strong advantage over conventional photovoltaics.

In addition to these, it will be necessary to determine whether there are any fundamental constraints peculiar to these systems which might limit their efficiency relative to other more proven systems, such as solid state PV devices.

Barriers to be overcome by biological systems for the production of hydrogen include the following.

- Suitable anaerobic organisms need to be selected which are not outgrown by competing organisms and whose hydrogen production, over a defined period of operation, does not decrease with time.

- A sufficient supply of carbon must be available in the growth media and this must be cheap and easily available. Utilising waste streams is potentially an attractive option, but under future more environmentally stringent conditions, the availability of waste may be reduced as waste production at source is minimised.

- Adequate light penetration into the reactor must be maintained to sustain microbial growth and hydrogen generation, but dense growth must not be permitted to reduce light penetration. Biofouling of the reactor surface must also be prevented to ensure adequate light penetration.

- Systems to separate and recycle the micro-organisms are needed to maintain an adequate active population within the system.

- Semi-continuous operation may well be required to balance the economics; this may introduce further specific requirements.

- The reactor temperature may need to be raised to aid the growth of those anaerobic bacteria that are relatively slow growing. Balancing this requirement with the need for maximum light penetration (and hence large surface area) may not be easy.

Institutional

A skills base both at the research and production/utilisation level is required in order to investigate the potential of these systems. There is currently no need for a manufacturing capability for full-scale systems. If any of these systems are to be developed for commercial use then appropriate investment backing will need to be found to move systems from the R&D phase to full-scale demonstration. In addition a user base for both system design and application would be needed.

As for small-scale conventional photovoltaic systems, an appropriate commercial environment for independent electricity generation will need to be established for ECPV cells.

For biological hydrogen systems, there are specific questions about future substrate availability. The backing of the water treatment industry and other similar concerns will be needed if the designs identified at the R&D level are to be progressed.

H. Prospects Modelling Results

Since resource-cost curves are not available for these technologies they have not been included in the MARKAL modelling exercise. No analyses of the possible uptakes of these technologies have been undertaken.

II. Programme Review

A. United Kingdom

Industry plays no role at the present time in the development of photoconversion in the UK. There is only a limited research community and the developments are still in their infancy at the fundamental research stage.

The Government's Programme

The UK Government funds fundamental research through the Research Councils, mainly based at universities. To date photoconversion has not been included in the DTI's Renewable Energy Programme, although a review has been undertaken to assess the potential of photoconversion as an energy production process in the UK. The review included both scientific assessment and engineering implications. Most research funds, both in the UK and internationally, are concentrated at the fundamental investigative level. There is currently no mechanism for taking the results of these basic studies forward to a pilot-scale 'demonstration of the theory'.

The Science and Engineering Research Council supports fundamental research including fundamental aspects of photoconversion. It has recently established a clean technology programme, in collaboration with the Agriculture and Food Research Council, which aims to support research into technologies which reduce the production of dangerous or unpleasant by-products in manufacturing processes. One of the themes they are promoting is harnessing photosynthesis to derive new chemical feedstocks and fuels. Support for fundamental generic studies is to be given to enable the potential of photosynthesis to be exploited. The logical next step for the technology is to move on from research to the demonstration of laboratory-scale systems.

B. Other Countries

Several countries have R&D programmes looking into various photoconversion systems. The highest level of activity is probably in the

USA. As a result of reviewing this overseas work it has been possible to narrow down the various technology options and identify electrochemical PV cells and photobiological hydrogen production as options worth investigating further in the UK. A conference sponsored by the International Energy Agency and the US Department of Energy in 1990 addressed the potential of photoconversion processes as a means of recycling carbon dioxide from the atmosphere. This meeting brought together scientists from thirteen nations and concluded that photoconversion processes could be harnessed to convert carbon dioxide to fuels and chemicals. It identified the need for more research at the fundamental level, very much in line with the aims of the UK's Clean Technology Programme initiated by the Science and Engineering Research Council.

Active Solar Heating Systems and Thermal Solar Power

I. Technology Review

A. Technology Status

Active solar systems convert solar radiation into <u>thermal energy</u> (heat) which can be used directly, stored for use in the future, or converted to electricity.

The UK climate has a <u>high fraction of diffuse solar radiation</u> and long periods of low radiation levels. Consequently, solar heating in the UK is best suited to lowtemperature heating applications, which do not require direct sunlight. The most common application is the provision of heat for domestic hot water (DHW) systems.

Figure 1 shows a common solar DHW system composed of a solar collector array, a preheat tank (optional), a pump, a control unit, connecting pipes, the normal hot water tank, and a conventional heat source. The conventional heat source is necessary because a standard solar system in the UK cannot provide sufficient heat to supply hot water at the desired temperature, throughout the year.

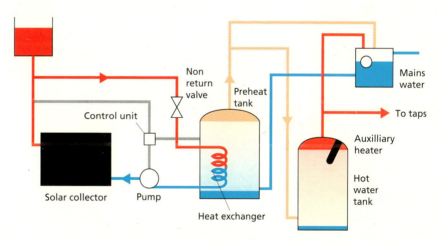

Figure 1: A typical active solar system.

In DHW systems, the solar collectors are usually mounted on the roof and provide heat to a fluid which is circulated between them and a water tank. If the collector is below the tank, there may be no need for a pump since the less dense heated fluid will be driven upwards by the more dense cold fluid. This leads to a natural circulation driven by gravity. Such systems are called thermosiphon systems. They are not common in the UK as they are incompatible with the design of most conventional DHW systems. More often a pump and control unit is used, where the pump is switched on only when solar heat is available and can be utilised.

The solar collectors fall into two basic categories.

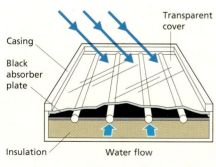

Figure 2: A flat plate collector.

- The flat plate collector (figure 2) has a blackened surface, this is the absorber plate which absorbs the incident solar energy. The energy is then transferred, as heat, to the fluid which usually flows through pipes connected to the plate, and then back to the tank. To reduce heat loss there is usually a glass cover on top and some form of insulation on the back.

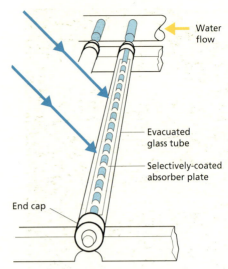

Figure 3: An evacuated tube collector.

• The alternative is the evacuated tube collector (figure 3). The principle is the same as for the flat plate collector, but each pipe and its absorber are sealed in an evacuated glass tube to reduce heat loss. Often the pipe contains a volatile liquid which evaporates on heating and condenses at the end of the pipe in a small heat exchanger. There the heat is passed to the fluid which carries it back to the hot water tank.

The fluid used to transfer the heat is often water with a nontoxic antifreeze additive for protection. It is possible to use the tap water itself but the risks of freezing make this uncommon in the UK.

In most DHW systems only the collectors are visible. They look similar to dark roof lights, but the overall area of the collector array is typically 4 to 5 sq.m, though they can range from 2 to 7 sq.m.

The other major use for the technology in the UK is for swimming pool heating, where it is particularly suited to pools used only between Spring and Autumn. These may be outdoor pools or enclosed pools where the air over the water is not conditioned. In these situations the pool water need only be a few degrees above ambient temperature. This allows the use of cheaper, unglazed collectors with less insulation. The pool water can be passed directly through the collectors, often using the filter pump (see figure 3), though for larger pools a more complex system is required. The collectors are situated near the pool, either on the ground, on a support structure or on a nearby building. A collector area equivalent to more than half the pool area is usually required.

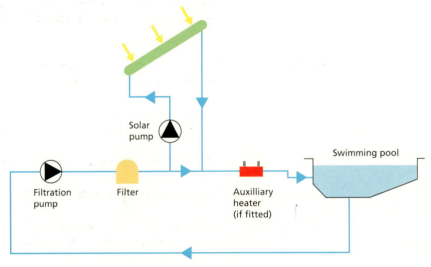

Figure 4: Typical solar heating system for a swimming pool.
(Courtesy of BSI Sales, Linford Wood, Milton Keynes).

Figure 5: Collectors on a house (courtesy of the Solar Trade Association).

Active solar heating systems can also be used to provide space heating in houses. However this is not common because large areas of collector (20 to 40 sq.m) are required to supply heat in winter when it is needed.

A solution to this problem is to collect the heat in summer and store it till winter. This can be achieved by storing the heat, usually in large water stores, and distributing it in district heating systems. These are called Solar Aided District Heating Systems (SADHS). A conventional source of 'top up' energy is usually required, although 70% or more can be supplied by the solar collectors. Such schemes have very large arrays of collectors.

Solar heat could also be used as low temperature industrial process heat.

Figure 6: Collectors on a swimming pool *(Courtesy Robinson's Developments).*

Finally, different sorts of collectors, which concentrate direct sunlight, can be used to produce very high temperatures. These can then be used either to produce electricity or for applications such as the detoxification of contaminated material. The climatic conditions in the UK appear to make this thermal solar power technology impractical here, since a large proportion of the incident solar radiation is diffuse.

B. Market Status

In 1991 there were over <u>twenty nine</u> firms involved in the UK active solar heating industry, as follows:

- Collector manufacturers 7+
- System installers 16+
- Collector importers 4+
- Controller suppliers 2+

Many of these are small commercial enterprises and for some solar energy is only a sideline to their main business (e.g. plumbing services). Most, but not all, are members of the Solar Trade Association Ltd.

Recent years have seen an increased interest in the technology and a consequent increase in production. In 1991, for instance, 36,000 sq.m of collectors were manufactured in the UK, an increase of 38% on 1990 production and 24,000 sq.m of these were exported; 1,000 sq.m of collectors were imported.

Of the collectors sold in the UK, more than 70% are used for domestic water heating (DHW) and most of the remaining 30% are used for swimming pool heating. In 1991 the total value of UK sales was £7.5 million, mostly for DHW systems.

The total value of active solar system sales world-wide in 1991 was approximately £520 million, of which £150 million was in Europe. In Northern European countries 60% to 70% are used for DHW, whereas in sunnier countries such as Greece, Israel, Japan and Australia the figure is nearer 80% to 90%. Most of the rest is used for swimming pool heating, though in Scandinavia a large number of collectors are used for Solar Aided District Heating Systems (SADHS).

In southern European countries the purchasers tend to come from the whole range of social classes and age groups. However, in northern European countries the purchasers tend to be mainly professionals motivated by concern for the environment. In the UK, the market is dominated by people who have just retired or are about to do so, since this is the period when they have most disposable income. They tend to live in the rural areas or affluent belts around towns. In countries like Germany, where professionals of all ages tend to have more disposable income, there is no such bias towards the older purchaser.

Current Status	Developed and deployed.
UK installed capacity (assumes all installations are still functional)	Domestic hot water: 36 GWh/year Swimming pool heating: 29 GWh/year Domestic space heating: almost none
UK industry	Small number of UK manufacturers: total industry approximately 29 firms

Table 1: Status of technology.

C. Resource

The potential for all forms of active solar heating is limited by the solar radiation available and the overall requirement for heat. The fact that direct sunshine is intermittent and unpredictable in the UK makes high temperature applications and electricity generation impractical, as mentioned in section A. However, for the other applications discussed, the high diffuse component is not inhibitive, and they are all practical possibilities.

Solar Domestic Hot Water
The Accessible Resource for solar domestic hot water systems by the year 2025 is estimated to be 12 TWh/year of which 9.6 TWh/year would replace oil or gas and 2.4 TWh/year electricity. This figure is primarily dependent on the number of suitable dwellings, the collector area and efficiency, and the hot water load.

A suitable dwelling should have a near south facing roof that is not significantly overshadowed. There is some evidence to suggest that this applies to around 50% of the existing housing stock in the UK.

For a given load, increasing the collector area also increases the solar energy supplied by the system. However, this is constrained by diminishing returns as the collector area increases. It is generally accepted that currently installed systems can supply up to 50% of the annual domestic requirement for hot water. Although more could certainly be achieved, it would usually not be worthwhile.

Statistics on domestic hot water load are sparse and not entirely consistent. The resource estimate is based on a predicted average household size of 2.4 persons by the year 2025 and a hot water consumption of 45 litres/person/day.

Most purchases of solar heating systems are not, at present, based on economic arguments. For many householders, the purchase of a solar heating system can be considered in the same way as the purchase of any other consumer durable. It is purchased because the householder wishes to own one for reasons of technical intrigue, to save the environment or fuel, or for general interest. Therefore, no economic considerations have been included in the Accessible Resource figures.

However, it must be remembered that the present take up is only a small fraction of the Accessible Resource. This is only likely to change significantly with a large increase in public awareness and the sale of more affordable systems.

Swimming Pool Heating

The main constraint on the potential resource for active solar heating for swimming pools is simply the number of suitable pools, primarily outdoor or enclosed pools. Currently, there are approximately 100,000 pools in the UK. Assuming this figure will roughly double by the year 2025, the estimated Accessible Resource by that time is 0.78 TWh/year. Of this 0.62 TWh/year would replace oil and gas and 0.16 TWh/year would replace electricity.

Solar Aided District Heating

A typical solar aided district heating system might include about 70% contribution from solar energy. Hence the main restriction on its potential is the availability of district heating schemes.

Such schemes could be retrofitted to existing housing estates, but they are likely to be most economic if incorporated into a new housing development, at the time of construction. It is therefore reasonable to limit the resource to new-build dwellings only. Further limitations include the percentage of dwellings built in clusters of a few hundred houses or more and the availability of land for the collector array and store. There is no data to indicate how restricting these factors are, so an applicability of 50% is assumed.

Given the following assumptions:

- a construction rate of 150,000 new dwellings per year;
- SADH systems fitted to 50% of all new dwellings from 1995 onwards;
- a heat load of 11,300 kWh_{th}/year/dwelling;
- a 70% solar contribution;

the Accessible Resource by 2025 would be 18 TWh_{th}/year.

However, district heating is not common in the UK and is not likely to experience a quick market uptake. Therefore the actual uptake of SADH is likely to be very small indeed, particularly when constraints such as the high capital cost and the long amortisation period are considered.

Industrial and Commercial Applications

Solar heating systems suitable for the UK only provide low grade heat which limits the potential for industrial process heat applications. The Accessible Resource for UK industry and commerce has been estimated at 2.9 TWh/year, though this may not be commercially viable.

D. Environmental Aspects

Active solar heating offers the advantages common to all renewable technologies in terms of displacing emissions from fossil-fuelled power production and avoids potential hazards associated with nuclear generation.

The only visible part of the system is the collector array. For domestic hot water (DHW) systems this is usually placed on the dwelling roof and is generally considered visually acceptable. For swimming pool applications it may be placed on a nearby building, on the ground around the pool or mounted on a frame nearby. Since the area used is between 50% and 80% of the pool area, a typical pool might have around 20 to 30 sq.m of collector array. This is unlikely to have a significant environmental impact.

For solar aided district heating systems (SADHS) the collector array could be quite large, maybe in excess of 6,000 sq.m. It would be mounted on frames in order to be inclined southwards at an optimal angle to the sun, and would therefore take up land equal to approximately twice its own area due to shadowing effects representing significant land use. The visual impact could be high. These considerations would have to be addressed individually for each scheme.

The environmental impact of collector manufacture is limited to that associated with the normal production of the component materials such as steel, copper, glass and insulation.

The environmental factors associated with active solar systems are summarised in table 2.

Environmental burden	Specification
Gaseous emissions:	None
Ionising radiation	None
Wastes:	None during use. Solid and Liquid waste on disposal.
Thermal emissions	None
Amenity & Comfort:	
Visual intrusion	Possibly significant for SADHS

Table 2: Environmental burdens.

Estimates of the emissions saved when using active solar systems in place of conventional generation systems are presented below.

Emission	Displaced emissions per kWh[a] in kg of oxide {b}, [c]	Annual savings for a scheme in an average household[d] (kg of oxide)
Carbon dioxide	0.443,{0.370},[0.734]	443, {370}, [734]
Sulphur dioxide	0.0036,{0.002}, [0.010]	3.6, {2}, [10]
Nitrogen oxides	0.0015,{0.001},[0.0034]	1.5,{1}, [3.4]

Table 3: Emission savings for a typical active solar domestic hot water heating system.
a: Approximately 80% of households employ gas or oil water heating and 20% use solid fuel or electricity. Many who use solid fuel in winter, use electricity for water heating in the summer months when solar systems make their greatest contribution. Therefore these figures are based on the 'average fuel mix' of 80% oil or gas and 20% electricity. For individual fuels the figures are given in brackets.
b : Figures in brackets give the values if the system replaces oil or gas water heating.
c: Figures in square brackets give the values if the system replaces electric water heating.
d: Based on average annual savings of 1,000 kWh/year, which might be expected for an 'average' household of 2.4 persons.

Emission	Displaced emissions (kg of oxide per kWh)	Annual savings for a typical private pool*(kg of oxide)
Carbon dioxide	0.370	2,200 to 3,300
Sulphur dioxide	0.002	12 to 18
Nitrogen oxides	0.001	6 to 9

Table 4: Emission savings for an active solar swimming pool heating system replacing either oil or gas heating. * Based on average annual savings of 300 kWh/year/sq.m of collector and assuming an average collector area of 20-30 sq.m.

The emission savings from swimming pool heating depends very much on the type of swimming pool being heated and the fuel being displaced. Table 4 gives the figures for an outdoor private pool where the fuel replaced is oil or gas. This is the largest relevant sector, with perhaps two thirds of the total potential in the swimming pool market.

A conventional district heating system would almost certainly use gas or oil as the heating fuel, so this is taken as the basis for the figures given in table 5.

Emission	Displaced emissions (kg of oxide per kWh)	Annual savings for a typical system *(kg of oxide)
Carbon dioxide	0.370	676,000
Sulphur dioxide	0.002	3,654
Nitrogen oxides	0.001	1,827

Table 5: Emission savings for a solar aided district heating system as compared to a conventional district heating system. *Based on annual savings of 1,827 MWh/year which may be expected from a system applied to 231 new houses in the UK.

Health

The possible growth of legionella bacteria needs to be considered for all domestic water heating systems, whether solar assisted or not. However, no case of legionnaires' disease that can be traced back to either conventional water systems in individual dwellings, or to solar water heating systems in buildings of any type, has been identified in the UK or overseas. It is generally thought that if a solar system is designed and installed in accordance with BS 5918:1989 and used to preheat water which is then heated to 60°C by a conventional water heating system, the risks of legionnaires' disease are no higher than those associated with the use of the conventional heating system alone.

E. Economics

The present day commercial price of a typical solar domestic hot water (DHW) system, varies greatly, as is shown in figure 7.

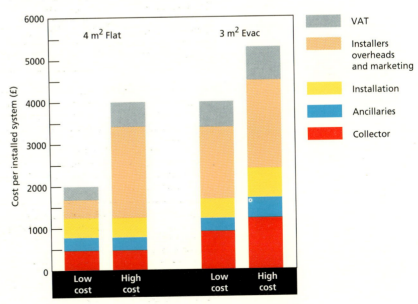

Figure 7: Examples of UK solar water heating system price breakdown (1991).

It is possible to buy systems which can be installed as a DIY kit. These can be considerably cheaper.

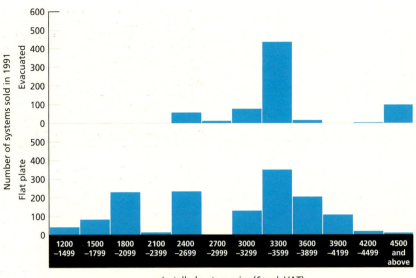

Figure 8: Prices for UK solar water heating systems (excluding VAT) 1991.

The great range in the price of typical systems is due mainly to the large variation in installers' overheads and marketing. Over recent years the task of explaining to the public how solar heating systems work, and demonstrating their effectiveness in the UK, has been borne almost entirely by the installers. This has significantly affected the final price. Often the highest price systems are sold by costly direct-mailing techniques. By contrast, low cost systems are generally sold by small companies who specialise in other business and only respond to specific requests from customers. Figure 8 shows the range of system prices, but it must be remembered that some of these are larger or smaller than the 'typical' system.

Typical system price (incl VAT)	£2,000 to £4,500
Typical system DIY price (incl VAT)	£1,000 to £2,500
Annual pump running costs	£6/year
Installation time	0.1 years
Assumed lifetime	25 years
Annual output	1,000 to 1,500 kWh$_{th}$
Distribution of energy output by season:	
Summer: June to August	39%
Autumn: September to November	24%
Spring: March to May	28%
Winter: December to February	9%

Table 6: Cost and performance data for a typical solar domestic hot water system.

In the existing UK market, purchasers do not buy for economic reasons alone. If an investment calculation is made, it is more likely to be based on simple payback methods (0% test discount rate) than on complex discount rate calculations. A simple payback calculation shows that the cost recovery period is likely to be over 10 years for a DIY system and over 20 years for a professionally installed system. It is worth noting, however, that there is the potential for significant price reductions should the mass market develop.

Typical system price (incl VAT) Includes DIY and professionally installed systems	£950 to £2,700
Annual running costs	Negligible
Installation time	0.1 years
Expected lifetime	15 years
Annual output (based on use from May to September)	~6,000 kWh$_{th}$

Table 7: Cost and performance data for a typical solar heating system for an outdoor swimming pool. Based on 20 sq.m of collector area.

Capital cost	~£1.34M
Annual recurring costs	~£65,000
Installation time	As for the district heating system
Expected lifetime	25 years
Annual output	1.8GWh$_{th}$

Table 8: Cost and performance data for the solar part of a solar aided district heating scheme. Based on figures for a scheme of 231 houses.

F. Resource-Cost Data

In the case of DHW systems the decision to purchase is not, at present, dependent primarily on cost. Since it is a consumer product rather than a technology investment it is not clear how to produce a meaningful resource-cost curve.

Similar arguments could be used for solar heated swimming pools. However, for outdoor or enclosed pools which are only used in the summer season, basic unglazed collector systems will almost always be economic and have a payback period of between 2 and 10 years.

The Maximum Practicable Resource by 2005 is likely to be about 1.75 TWh/year for DHW and 0.12 TWh/year for swimming pools at a cost of less than 10p/kWh using an 8% discount rate. The main factor influencing these figures is the development of sufficient production capacity.

A resource cost curve for SADH systems would be a complex matter with critical dependence on the market uptake of basic district heating systems. At present there is insufficient knowledge of the barriers to the uptake of district heating systems in general to undertake this task.

On the assumption that all new domestic buildings suitable for the inclusion of solar aided district heating schemes, have them installed, the Accessible Resource by 2005 would be 5.9 TWh/year. Since such schemes should be able to produce heat at less than 10p/kWh, the Maximum Practicable Resource is equal to this figure. In this analysis no account is taken of the market barriers to district heating schemes in general.

The Maximum Practicable Resource for industrial applications of active solar systems has not received detailed assessment because in most situations there is the opportunity to recover heat from other processes more cheaply.

G. Constraints and Opportunities

At the present time market issues rather than any technical issues are perceived to be the major barriers to increased adoption of solar domestic hot water (DHW) systems. Solar heating does give individuals the opportunity to make a visible personal commitment to renewable energy.

Technological / System

If an increase in production levels becomes feasible, either for the home market or for exports, there might be the possibility of reducing costs by investment in production engineering techniques. An increased market could also justify the redesign of some systems for easier installation. This could help reduce costs still further by limiting the installation time. It could also help stimulate a greater DIY market.

Solar swimming pool heating is already easily available and economically attractive. The main barriers to its greater uptake are probably lack of awareness and perceived visual intrusion.

The uptake of solar aided district heating systems is linked to the uptake of small-scale district heating systems. When seen as an input to a "conventional" district heating scheme, the prospect of providing a solar energy input could, if the price were right, enhance the commercial attractiveness of such a scheme. This could influence the potential for district heating as a whole. Nevertheless, there are considerable barriers to its uptake including the high capital investment required and the long cost recovery period. A detailed assessment of the potential and constraints for district heating in general in the UK is beyond the scope of this module.

Institutional

Experience in other countries indicates that the market for this technology will take off when system prices reach the point where they are readily affordable by the average householder, and where they make solar heating more economically attractive than it is in the UK today.

Figure 7 shows that a substantial proportion of the cost of a domestic active solar water heating system goes towards installers' overheads and marketing. This is necessary because of a general lack of public awareness and confidence in the technology. If the market achieves a substantial expansion then the unit marketing costs are likely to fall. This necessarily would be preceded by increased public awareness and confidence in the technology, possibly encouraged by more information/promotional material. This is perhaps the most important opportunity for increasing the market uptake of these systems.

An additional constraint is the fact that this is a heat-producing technology. It is therefore more difficult to get support for it than for the electricity generating technologies.

In the past there has also been a further constraint in that VAT has been levied on active solar systems while it has not been applied to domestic fuel. The recent proposal to apply VAT to domestic fuel, will, however change this situation.

The need to prevent the growth of micro-organisms in hot water systems is imposing more stringent operation and maintenance requirements on the owners and operators of non-domestic buildings. The incorporation of active solar heating systems into the hot water supply of such buildings could impose an additional maintenance burden; this may constrain the deployment of this technology in these building types.

H. Prospects Modelling Results

The results of the modelling analysis show that in the particular conditions portrayed by the "Heightened Environmental Concern" scenarios, there is a significant potential for energy production by active solar technologies. Indeed in 2025 there is a predicted uptake of 2.53 TWh/year.

The environmental sensitivity studies and Shifting Sands scenario do not predict any change from the base case active solar uptakes.

Scenario	Discount rate	Contribution, TWh/year			
		2000	2005	2010	2025
HOP	8%	0.02	0.01	0.01	0
	15%	0.02	0.01	0.01	0
	Survey (10%)	0.02	0.01	0.01	0
CSS	8%	0.02	0.01	0.01	0.01
	15%	0.02	0.01	0.01	0
	Survey (10%)	0.02	0.01	0.01	0
LOP	8%	0.02	0.01	0.01	0
	15%	0.02	0.01	0.01	0
	Survey (10%)	0.02	0.01	0.01	0
HECa	8%	0.02	0.01	0.01	2.5
	15%	0.02	0.01	0.01	0
	Survey (10%)	0.02	0.01	0.01	0
HECb	8%	0.02	0.01	0.01	2.5
	15%	0.02	0.01	0.01	0
	Survey (10%)	0.02	0.01	0.01	0

Table 9: Potential contribution under all scenarios.

In most scenarios the use of active solar for domestic water heating gradually falls away to zero contribution by about 2015. This is because the present cost of the systems is relatively high and the model predicts that the energy can be supplied more cheaply by other methods. Working on the basis that no new systems will therefore be installed, the contribution gradually falls away from the present value of about 0.03 TWh/year to zero as the current systems reach the end of their useful life. However, in the 8% CSS and HEC scenarios, the contribution falls off towards about 2015, then picks up again towards 2030. In the CSS scenario the pickup is slow, and only catches up with the present utilisation of the technology between 2025 and 2030. It reaches 0.05 TWh/year in 2030. This pick up is probably due to the increase in prices of alternative fuels which would make active solar more favourable in certain specific situations. In the HEC scenario the uptake is somewhat larger and reaches a contribution of 2.53 TWh/year by 2025. The reason is that with Heightened Environmental Concern there is likely to be greater publicity of active solar. This will lead to greater public awareness which could reduce the need for the present expensive forms of marketing and thus reduce the price of systems by up to 50%. This reduction in price is likely to be the main reason for the larger predicted uptake in the HEC scenarios.

The predicted fall off in the application of active solar hot water heating in the next 20 years is probably unrealistic. It is based only on economic considerations, which would also lead to the conclusion that there should be no current use of these systems. However this is not the case: the solar domestic hot water contribution is currently running at about 0.03 TWh/year. The reason is that most purchasers of active solar systems do not buy on the grounds of economics, but rather for reasons of environmental concern, technical intrigue etc. It is likely, therefore, that this market will continue at least at its current level.

II. Programme Review

A. United Kingdom

There is already a UK industry based around the manufacture and installation of active solar systems for providing domestic hot water. Currently around 35,000 sq.m of collectors are manufactured in the UK each year and of these about 13,000 sq.m are sold to the home market. A major manufacturer of evacuated tube collectors is based in the country and one of the Regional Electricity Companies (RECs) has now started marketing active solar systems.

The main barrier to the widespread uptake of these systems is their cost. The largest part of the cost of a system goes on installers' overheads and marketing. This is the area where there is greatest potential for reduction. An increase in support for publicity on this technology would help improve public awareness and confidence, and thereby open up the opportunities for cost reductions. It has also been suggested that VAT could be removed from these systems because of their environmental benefit. This would certainly help to stimulate the market. If a mass market begins to develop there would be an opportunity to reduce costs still further by manufacturing economies of scale.

The Government's Programme
The Government's Active Solar R&D Programme, carried out by the former Department of Energy between 1977 and 1984, coincided with a period of expanding world demand for active solar products. The Programme was designed to define the potential contribution of the technology to UK energy supplies and to stimulate the development of cost-effective systems.

Although several of the 70 or so projects examined other applications, the programme concentrated on the application of active solar heating to domestic water and space heating. Particular attention was paid to component and system development, laboratory testing, field trials and modelling studies.

The programme proved that active solar heating is technologically viable in the UK, particularly in low temperature applications.

Since 1984 the Government maintained a watching brief in the area until a recent review was undertaken, completed in 1992. This review predicted a potential resource of around 12 TWh/year for solar domestic hot water heating alone. However, to encourage a mass market to develop it concludes that it will be important to raise public confidence in the systems by providing appropriate information and promotional material.

B. Other Countries

The world-wide sales of active solar systems are worth approximately £500 million per year, of which about 4% are for swimming pool applications and most of the remainder are for domestic water heating systems. Europe represents about 30% of the world market. The largest national markets are in the USA, Japan, Israel, Australia and Greece. There is also a substantial but unusual market in Austria where installation of DIY systems has been encouraged by many consumer groups.

In sunny countries such as Greece, Israel, Japan and Australia, 80 to 90% of collectors are used for domestic water heating. In northern European countries the figure is slightly less at 60 to 70% with more collectors being used for swimming pool heating. There are also a few sales of systems to provide hot water in industrial, commercial and hospital buildings. In Scandinavia a significant proportion of collectors is used for large-scale solar-aided district heating schemes and in Germany there is an interest in solar space heating. Very few commercial collectors are used for other applications although there are examples of home-built air heating collectors being used for hay drying in Finland, Sweden, Switzerland and France.

Active Solar Heating Systems and Thermal Solar Power

Passive Solar Design

I. Technology Review

A. Technology Status

Passive solar design (PSD) aims to maximise the benefit of free solar gains to buildings. Properly applied PSD can reduce the energy needed for space conditioning (heating, cooling and ventilation) and lighting while still maintaining acceptable internal environmental conditions. PSD is most effective when combined with energy efficiency measures in an integrated approach to energy conscious design. Solar energy technologies depend upon the availability of sunlight, and unfortunately, this is a variable resource in temperate maritime countries like the UK. Extensive periods of cloud-free direct sunlight are not required, however, for simple PSD measures. One of the key applications – to provide high levels of glare-free natural lighting – is in fact more easily achieved under overcast conditions.

Figure 1: Passive solar estate in South London.

PSD is not a new technology. Modifying the design of buildings to encourage natural ventilation or to let in more daylight has been common practice since the beginnings of architectural science. Usually these modifications were made for aesthetic or health reasons. People prefer daylight and appreciate the other benefits of windows such as view, sunlight availability and natural ventilation. It is only recently that the potential energy benefits from the wider use of PSD have been appreciated. The basic principles behind PSD are described as follows.

Heating
- Orienting buildings and distributing glazing in a way that allows the interior to be heated by solar radiation, while at the same time minimising heat losses from shaded facades and maintaining thermal comfort for the occupants.

Lighting
- Arranging vertical and horizontal glazing to allow the utilisation of daylight thus reducing the need for artificial lighting while maintaining visual comfort.

Cooling
- Allowing solar heated air to assist natural convection within a building, thus reducing or obviating the need for mechanical ventilation and cooling systems.

For these measures to be effective buildings should be arranged to minimise overshading and provide protection from adverse micro climate effects. It is also important to design the internal layout of buildings so that the spaces receiving the solar benefits are occupied and use is made of the improved environment. In houses, for example, this means locating the rooms with the highest temperature requirements – living rooms and main bedrooms – on the south side, and circulation spaces and areas with lower heating requirements – bathrooms etc. – on the north side. Good design can minimise any potential for overheating during occupied periods. Designing for daylighting involves the choice of appropriate shading devices to reduce the risks of glare and overheating. Innovative daylight systems such as light shelves, light pipes, mirrored louvres and holographic or prismatic glazing can be used to improve the uniformity of daylight. Atria or courtyards can be used as secondary light sources. Daylight-linked lighting controls can ensure that electric lighting is displaced; these controls include, but are not limited to, automatic photoelectric (i.e. light sensitive) devices.

More complex approaches have been developed (atria, Trombe walls etc.) but these are essentially more sophisticated applications of the same principles.

The applicability of PSD and the scale of energy savings achieved through it depends upon the building type. The space heating requirements of individual houses can be reduced by around 1,000 kWh/year through the adoption of simple PSD measures. More complex measures can increase this to about 2,000 kWh/year; however not all these will be cost-effective in every application. Although modest on an individual unit basis, when scaled up across the national housing market the total energy and carbon dioxide savings can become significant. Solar-driven ventilation and daylighting are not generally significant in housing since houses are small enough to allow natural ventilation through window openings and artificial lighting is usually only required in the evening. During the hours when it is available, daylight remains the preferred light source in housing. In their homes people already tend to make the most of daylight by switching off electric lighting when daylight is sufficient.

If PSD is to be efficiently exploited in individual houses, PSD principles must also be applied to the design and lay-out of whole housing estates. A passive solar housing estate design must allow for unshaded solar access and correct orientation for as many of the houses as possible so that each individual house can gain the maximum energy benefit. This can require innovative design to enable acceptable street access, good visual appearance, privacy and so on. If planning policy is too prescriptive in terms of road layout, dwelling spacing etc., the achievable solar benefits could be limited. It will usually be necessary to compromise the estate design to include a mixture of passive and non-passive houses.

Non-domestic buildings – offices, schools, factories etc. can obtain significant individual energy benefits from PSD. Usually the most effective PSD measure in these building types, at least from an energy cost and carbon dioxide savings point of view, is daylighting. This displaces peak rate electricity and can therefore produce high cost benefits from modest energy savings if premises are designed to maximise the available daylight. Most of the daylight savings predicted come from the application of daylight-linked lighting control in new and refurbished buildings. In new buildings and through the recladding of existing ones, a further, but smaller, contribution can be achieved by optimising the size of windows. The provision of natural cooling mechanisms is another important PSD measure in non-domestic buildings. The possible savings (from heating, lighting and cooling measures) per unit are dependant upon the size and use but can be in the order of 20,000 to 30,000 kWh/year for a typical unit. However, it is often impossible to achieve the maximum benefits cost-effectively.

Some PSD measures such as conservatories and roof space collectors can be retrofitted. This is usually best considered when refurbishment or extensions are being planned. Modifying glazing design in non-domestic buildings during refurbishment is a major opportunity for incorporating daylighting design considerations.

B. Market Status

PSD is a commercially available technology. A small but growing number of architectural practices have experience in applying PSD measures, particularly in housing. This is true in the UK, in Europe and in the USA. The 'innovator' market has therefore been established and must now be expanded into the 'imitator' stage of market growth.

It is difficult to quantify the size of the installed capacity resulting from PSD in the UK and overseas. All buildings use solar energy to a greater or lesser extent; the unplanned use of sunlight in buildings makes significant use of this energy. It is estimated that this adventitious use of solar energy is making a contribution to space heating and lighting in UK buildings of approximately 145 TWh/year. Buildings deliberately designed for enhanced use of solar energy are still few in number. Only some 500 houses and 100 non-domestic buildings have been constructed in the UK with a specific PSD element.

Climatic variations affect the application of PSD. There is more to be gained from solar-driven ventilation and cooling in warm climates and more to be gained from solar space heating in temperate areas. Daylighting benefits are largely climate independent. The variation in climate across the UK is not large enough to have any significant effect on PSD strategies, but there will be some variation across the country in the way the measures are applied. In Scotland, for example, the longer season for which heating is required can be countered by making use of the longer hours of daylight available in late spring and early autumn.

Figure 2: South Staffordshire Water Company offices which make use of daylighting as a passive solar measure.

A summary of the status of the technology is given in table 1.

Current status	Housing: demonstration Non-domestic: development/demonstration
UK installed capacity:	~ 145 TWh/year – from all buildings ~ 1.5 GWh/year – from specifically designed buildings
Europe: installed capacity:	~ 1,200 TWh/year – from all buildings
UK industry	Small number of designers and builders with experience of passive solar design

Table 1: Status of technology.

C. Resource

There are two important influences on the Passive Solar resource in the UK, firstly the availability of solar energy and secondly the rate of newbuild and refurbishment in the building market. The solar radiation pattern over the seasons is well documented. From these statistics the energy saving potential from PSD can be estimated with reasonable accuracy. New-build and refurbishment rates are variable and determined by economic, social and political factors. There is a continuing need to upgrade and replace the building stock and so new opportunities for PSD arise continuously. However, if the opportunity to include some PSD measures is missed at the design stage of a building project then that opportunity is lost and may not recur again for some 60 years or more - the lifetime of the building.

The major constraints on the increased use of PSD are:

- the degree of solar utilisation compatible with other requirements on buildings;

- specific site constraints;

- overall level of building activity;

- level of knowledge and expertise in the building professions;

- the perceived risk by a conservative building industry;

- the general barriers to energy efficiency in buildings (payback etc).

The resource for PSD depends on the opportunities for its use created by newbuild and refurbishment of existing buildings. Since this fluctuates from year to year, average values for building rates have been used where available, based upon DoE data for recent years. An Accessible Resource has been calculated on the assumption that all newbuild and refurbishment will benefit from PSD from now until the year 2025. In reality, of course, PSD would need to penetrate the industry sufficient to build up a pool of knowledge and expertise and stimulate market demand. Thus to assess the Maximum Practicable Resource, a simple market penetration model has been used. This assumes a linear penetration of the market of 2% of each year's new build, up to a maximum of 30%.

Considering the exploitation of daylight in isolation, the conventional use of daylighting (i.e. the provision of windows) is already widely deployed. It is possible to try and identify extra savings from taking active measures to exploit daylight. There are a number of exemplar buildings in the UK where this has been done. Of the more innovative systems, mirrored louvres and prismatic glazing are now commercially available in the UK, although currently only one building has prismatic glazing and none have mirrored louvres. There are however a number of buildings in Western Europe e.g. The Netherlands and West Germany where these techniques have been used. A small number of building in the UK has light shelves fitted. Because of the uncertainty over the size of energy benefits, the cost-effectiveness of such systems cannot be estimated.

England & Wales	Scotland	Northern Ireland	UK
8.5 TWh/year	1 TWh/year	0.5 TWh/year	10 TWh/year

Table 2: Accessible Resource in the UK, estimated primary energy savings for buildings incorporating a specific PSD intent. Note that these figures are based upon estimates of future build rates which are subject to considerable uncertainty both in magnitude and regional variation. The resource estimated here is only for simple measures which add little or no cost to a design. The resource is therefore independent of discount rate.

The current lighting controls market is estimated at £10 M/year, roughly corresponding to 4 sq.km of non-domestic floorspace. However, only around a half of this floorspace is estimated to be daylight linked. The total floor area of the non-domestic sector is estimated to be 970 sq.km.

An estimate of the potential energy savings from daylighting design is shown in table 3.

Building Type	Primary energy consumed for lighting (TWh per year)	Range of Accessible Resource from daylighting design by 2020 (TWh primary energy per year)
UK non-domestic buildings – offices, education, factories, warehouses, health service and multi-residential	44.8	5.6 to 11.2

Table 3: Estimated Accessible Resource in UK from daylighting design measures.

Note that the range of predicted energy savings is based upon the Building Research Establishment (BRE) estimates of percentage lighting energy savings for different building types. The numbers are not directly comparable with those in table 2 but they clearly indicate the importance of daylighting as a PSD measure.

The provision of natural cooling mechanisms using PSD measures is also thought to have the potential for significant energy savings but it is currently difficult to quantify these savings.

D. Environmental Aspects

Providing that adequate attention is paid to good design to avoid such problems as glare and overheating – which could result in additional energy loads through increased use of artificial light and air conditioning – PSD introduces no emissions to the environment additional to those associated with the normal production of building materials and components. Apart from its energy saving potential, PSD can bring other benefits. For example through the exploitation of daylight fewer lamps could be used, and daylighting design which results in a narrow plan building or use of courtyards/atria can help avoid air conditioning. This in turn would reduce associated emissions from cooling plant.

When applied to new buildings, PSD necessarily has land-use implications. The encroachment of new developments into the green belt areas is a matter of concern and is subject to local structure plan legislation etc. The most extensive application of PSD measures to date has been in green field new town developments, particularly Milton Keynes. The construction of further new towns of this kind is the subject of much debate at the present time. Housing estate developments employing PSD measures are not necessarily more land hungry; work has shown that developments of up to 40 dwellings per hectare are technically possible. It is true however, that more relaxed densities provide more scope for maximising PSD benefits.

The adoption of PSD in newbuild or as retrofit has some visual consequences. Asymmetric glazing distributions on houses, attached glazed spaces and possible lack of privacy are some examples. Passive Solar designs can be made aesthetically attractive but architectural mistakes can also be made. On the other hand, it is important to recognise that PSD can result in enhanced environmental conditions. Occupants can reap the benefits of more comfortable interiors, and the external architecture of houses, estates and commercial developments can be much improved as a consequence of its use. PSD can also be used as a way of improving the amenity of buildings and avoiding health problems associated with building designs less well attuned to prevailing climatic conditions. There is a growing body of evidence that the occupants of buildings – both residential and non-residential – prefer light and airy internal conditions and attach a high priority to having a view – provided that effects such as glare and overheating can be prevented or controlled.

The environmental factors associated with PSD are summarised in table 4.

Environmental burden	Specification
Gaseous emissions	None
Ionising radiation	None
Wastes	Normal building wastes
Thermal emissions	None
Amenity/Comfort:	
Visual intrusion	Some impact possible
Use of wilderness area	Some impact possible

Table 4: Environmental burdens associated with passive solar design.

The carbon dioxide reductions accruing from PSD are estimated in table 5. These figures assume that the displaced fuel for houses is natural gas and for non-domestic buildings is two thirds electricity and one third natural gas.

Emission	Displaced emissions (grammes of oxide per kWh)		Annual savings per typical scheme (tonnes of oxide per year)	
	Houses	Non-domestic	Houses	Non-domestic
Carbon dioxide	370	610	0.37	18.5
Sulphur dioxide	2	7.3	0.002	0.25
Nitrogen oxides	1	2.6	0.001	0.1

Table 5: Displaced emissions arising from PSD. For houses these figures assume that PSD is displacing energy provided by a gas boiler. For non-domestic buildings the figures assume similar gas boiler savings as well as power station savings resulting from reduced electrical demand for lighting and ventilation.

Health

There is increasing evidence that high levels of natural lighting and natural ventilation can significantly improve the working environment inside buildings. PSD, by its very nature, can be an effective way of achieving naturally conditioned buildings.

E. Economics

PSD, unlike most other renewables, does not require plant or other capital-intensive investment. The simplest PSD measures can often be introduced into a design at little or no identifiable extra cost. Some measures can have a cost element, e.g. increased areas of high performance glazing or a lighting control system to maximise use of daylight. The cost-effectiveness of these measures is usually judged on a simple payback calculation for the consumer rather than on a discounted cash flow analysis. The lifetime is generally that of the building i.e. 60 to 100 years. Some refurbishment is, however, likely on a shorter time scale, particularly for non-domestic buildings. Acceptable payback periods are usually much shorter depending on circumstances, but for most energy efficiency measures they are in the range 1 to 10 years.

PSD measures are therefore normally assessed on their marginal costs and benefits. A simple measure such as rearranging the glazing distribution on a house can result in energy benefits at no extra cost. More complex design approaches, for example maximising the daylighting in an office which may demand the provision of shading devices for control of overheating and glare, will have an identifiable additional cost. The magnitude of the energy benefits must be sufficient to justify such extra costs. The time scales over which the costs are amortised and the savings aggregated are subject to the investment criteria of the building developer or owner. A significant problem in the commercial building sector is that many buildings are not owner occupied, so the energy benefits are not realised by the organisation paying the construction costs. There is, therefore, little incentive to bear additional costs due to PSD or other energy efficiency measures unless they can be shown to enhance the marketability of the building.

A sample calculation is given below for the installation of daylight-linked lighting control. For this particular measure, costs need to be compared with a reference control system (usually a row of manual switches) which would need to be provided anyway. Typical extra costs range from zero to £2/sq.m of floorspace. These figures have been used in table 6.

Cost/sq.m		£0	£2
Typical delivered energy saving kWh/sq.m/year	4.3		
Equivalent annual cost at 8% discount rate	per kWh saved	£0	£0.05
Equivalent annual cost at 15% discount rate	per kWh saved	£0	£0.08
Equivalent annual cost at 25% discount rate	per kWh saved	£0	£0.12
Lifetime of measure (years)	15		
Fuel saved	Electricity		

Table 6. Daylighting financial assessment – non-domestic sector

F. Resource-Cost Data

Since PSD is not a power generation technology, resource-cost curves of the kind used in other modules are not appropriate. The potential contribution from PSD can be quantified from an analysis of the unit energy savings appropriate to different building types and market sectors, future building construction rates and likely levels of penetration of the technology. A basic set of data has been prepared to represent simple PSD measures which can be assumed to have no investment cost associated with them, e.g. orientation and glazing distribution. This data has been used for the Composite (CSS), Low Oil Price (LOP), High Oil Price (HOP) and Shifting Sands (SS) scenarios in the MARKAL modelling analyses. A range of additional, more advanced, measures can be proposed which have energy saving potential but carry an identifiable investment cost, e.g. roof space collectors. This additional data has been used for the Heightened Environmental Concern (HEC) MARKAL scenarios. The data inputs are summarised in tables 7 and 8 for the two sets of scenarios.

Housing:	Newbuild
Unit energy saving (kWh/year)	1,000
Build rate (units/year)	150,000
% suitable for PSD	50
Penetration rate (%/year)	2
Penetration limit (%)	30

Industry:	Newbuild	Refurbishment
Unit energy saving (kWh/sq.m/year):		
heating	1	1
cooling	1	0
lighting	5	2
Build rate (sq.m/year)	5,200,000	4,200,000
Penetration rate (%/year)	2	0.5
Penetration limit (%)	30	7.5

Commercial:	Private		Public	
	newbuild	refurbishment	newbuild	refurbishment
Unit energy saving (kWh/sq.m/year):				
heating	2	1	2	4
cooling	1	0	1	0
lighting	6	2	4	1
Build rate (sq.m/year)	1,600,000	550,000	400,000	1,370,000
Penetration rate (%/year)	2	0.5	2	0.5
Penetration limit (%)	30	7.5	20	5

Table 7: PSD input data for MARKAL CSS, HOP, LOP and SS scenarios.

Housing:	Newbuild
Unit energy saving (kWh/year)	1,000 *(500)*
Build rate (units/year)	150,000
% suitable for PSD	50
Penetration rate (%/year)	2
Penetration limit (%)	100

Table 8 continued overleaf

Industry:	Newbuild	Refurbishment
Unit energy saving (kWh/sq.m/year):		
heating	1	1
cooling	1 *(1)*	0 *(1)*
lighting	5 *(1)*	2 *(1)*
Build rate (sq.m/year)	5,200,000	4,200,000
Penetration rate (%/year)	2	0.5
Penetration limit (%)	100	50

Commercial:	Private		Public	
	newbuild	refurbishment	newbuild	refurbishment
Unit energy saving (kWh/sq.m/year):				
heating	2	1	2	4
cooling	1 *(1)*	0 *(1)*	1 *(1)*	0 *(1)*
lighting	6 *(1)*	2 *(1)*	4 *(1)*	1 *(1)*
Build rate (sq.m/year)	1,600,000	550,000	400,000	1,370,000
Penetration rate (%/year)	2	0.5	2	0.5
Penetration limit (%)	100	50	100	50

Table 8: PSD input data for MARKAL HEC scenarios.

Figures in brackets show the incremental energy saving measures for the HEC scenarios. It is assumed that these will become available five years after the basic measures. The investment cost associated with these measures has been set at £0.18/kWh of incremental energy saving over the life of the measure. The lifetime for all PSD measures is assumed to be 60 years. This level of investment cost is equivalent to that used by MARKAL for double glazing.

The additional savings that may be attainable from the wide spread application of daylight design measures have not been modelled by MARKAL.

G. Constraints and Opportunities

Technological / System
The application of PSD does not, in general, depend upon major developments in the performance or cost of the technology. The main barriers to uptake are therefore institutional.

There are, however, some areas where technological developments will benefit the technology – for example, the development of more responsive heating and lighting control systems to derive optimum benefit from solar heating and daylighting gains. An increased understanding of the mechanisms of natural ventilation in buildings is also needed. Specifically, we need to know how solar gains can influence and enhance these mechanisms.

Developments in glazing systems can be beneficial to PSD since they can make design solutions more tolerant to increased glazing areas without incurring penalties of increased heat loss or overheating. Over the last decade, low emissivity glazing has been developed by the industry so that it is now widely available commercially. More advanced glazing technologies – electrochromic systems etc.– are now under development. These could have an important influence on conventional, and especially passive solar, design. However, these systems need to undergo considerable development and testing to prove their performance and reduce their costs.

Similar opportunities exist for transparent insulation materials and systems. The simple forms and applications of these materials are being tried in the market now. The energy benefits are limited in highly insulated new dwellings but may have larger benefits in refurbished structures. Important experience is being gained, and more sophisticated forms are under development, which may have more potential for energy savings in a wider range of building types.

An equally important technical development is the establishment of a skills base in the design and construction professions to make PSD easier to implement. This requires the provision of technical information and guidance suitable for a range of uses, for example in schools of architecture, by experts giving specific site and project guidance. This information and guidance needs to be properly underpinned with technical backup and accredited as authoritative and tested.

Institutional
The main institutional barriers to the uptake of PSD are discussed below.

Decisions to embrace PSD as an energy-saving device can only be made when the decision-makers involved in a project understand exactly what is involved in the adoption of the technology. Results from market studies indicate that lack of awareness is widespread within the industry and among the general public. When designers do choose to consult information, they frequently turn to information supplied by product manufacturers. But few manufacturers make products specifically designed to exploit passive solar gain in the form of daylight. Daylighting is not a design innovation which many manufacturers seek to promote. Even where awareness is slightly higher, as it is among some of the design professions, there are likely to be misapprehensions about the costs and benefits of solar design in particular building types, and about how such buildings are designed. Awareness is not, of course, the same as the ability to implement PSD.

There are many perceived risks of adopting a solar design approach which could hinder its more widespread adoption. In the non-domestic buildings sector such risks might include:

- lack of examples of the technology and authoritative information;

- lack of confidence in the energy, cost and amenity performance achievable;

- unsuitable investment and funding criteria.

For housing such perceived risks include:

- lower site densities;

- less saleable houses;

- summer overheating.

In addition there are some designers who tend to be hostile to the notion of exploiting daylight – mainly because they see glazing as a weak thermal link in a building's external envelope.

In reality, many of these apparent risks may stem from insufficient information on the true benefits and costs of applying the technology. For example, it is currently a widespread view that passive solar buildings are more expensive to build than ordinary buildings, yet current design and monitoring studies indicate that this is not necessarily the case.

The most appropriate and credible means of getting design guidance to the right members of the design teams is an issue to be considered. This is made much more complex by the tendency of the design community to resist stereotyping in the way they assimilate information. Allowance must be made for this diversity by pursuing various types of design support. Examples of these options are:

- technology transfer activities to address the skills shortage by providing specific energy design advice to building projects;

- design and monitoring case studies: results from technical projects will produce case study material to inform designers;

- interactive media – electronic systems for the storage and retrieval of visual data can provide complex information in a usable form;

- appropriately packaged design tools – these can help incorporate the necessary numerical analysis of, for example, daylight factors or peak temperatures into the design process.

The complex structure of the building industry with its plethora of professional, institutional and other bodies, creates a situation in which many parties share the responsibility for design and construction decisions. Each of these parties regards their input to the building process as a specialism in its own right, rather than as part of an integrated approach to design issues. In addition, the contractual arrangements in some of the professions can act against PSD.

At a more pragmatic level, the impact which PSD can have on energy use in buildings is determined by the degree of activity in the building sector. This is particularly true for new-build measures which offer the easiest technical route for PSD. At times when the market is shrinking, there is little opportunity or incentive for innovative schemes. At times of very high activity, the industry can become overloaded and not have the 'thinking time' to consider new measures.

Probably the most effective means of accelerating the uptake of PSD is through its recognition in building regulations and through the incorporation of that recognition into professional codes of practice. This would initially substitute for market pull and would generally raise awareness of the technology. However, issues such as perceived risk and the development of design aids would still need to be addressed. A related approach is through building energy labelling and other assessment schemes, such as environmental audits. Some schemes of this kind are now in operation in the market. They are attractive for PSD because they must be applied at the design stage of a project when there is still time to incorporate changes. Here it will be necessary to ensure that the assessment procedures take full account of passive solar benefits.

H. Prospects – Modelling Results

The MARKAL modelling analyses have provided predicted uptakes of PSD in the domestic sector only. The results have therefore been manually scaled (by comparison with the input data for the other market sectors given in section F) to obtain predicted uptakes for PSD in all sectors. The figures only include new-build PSD and a minimal estimate for refurbishment.

They are summarised in table 9.

Scenario	Discount rate	Contribution, TWh/year			
		2000	2005	2010	2025
HOP	8%	0.16	0.31	0.52	1.1
	15%	0.16	0.31	0.52	1.1
	Survey (10%)	0.16	0.31	0.52	1.1
CSS	8%	0.16	0.31	0.52	1.1
	15%	0.16	0.31	0.52	1.1
	Survey (10%)	0.16	0.31	0.52	1.1
LOP	8%	0.16	0.31	0.52	1.1
	15%	0.16	0.31	0.52	1.1
	Survey (10%)	0.16	0.31	0.52	1.1
HECa	8%	0.22	0.69	1.5	4.3
	15%	0.22	0.69	1.5	4.3
	Survey (10%)	0.22	0.69	1.5	4.3
HECb	8%	0.22	0.69	1.5	4.3
	15%	0.22	0.69	1.5	4.3
	Survey (10%)	0.22	0.69	1.5	4.3

Table 9: Potential contribution under all scenarios.

The results indicate a steadily increasing uptake of PSD over time. The CSS, HOP & LOP results are all the same, similarly the HEC scenarios are identical.

The secondary energy savings resulting from the uptake of PSD predicted by MARKAL clearly reflect the input data assumptions about market penetration. The uptakes are insensitive to discount rate since for the majority of them there is no investment cost to be taken into account. The HEC results indicate that the modest-cost additional measures are adopted for all three discount rates.

If these energy savings are compared with the estimated Accessible Resource for the UK of 10 TWh/year (primary) then under the HEC scenarios the 2025 predicted uptake of 4.28 (8.75 primary) TWh/year amounts to about 87% of this resource. Under the CSS scenarios the 2025 uptake of 1.1 (2.3 primary) TWh/year amounts to 23% of the resource.

These energy savings result from a conservative estimate of the application of PSD measures in refurbishment. An upper limit for the energy savings resulting from a broader application of PSD in refurbishment has been estimated (outside of MARKAL) at 5.8 TWh/year (delivered).

The environmental sensitivity studies and Shifting Sands scenario do not predict any change from the base case PSD uptakes.

II. Programme Review

A. United Kingdom

Passive solar design, in its simplest forms, is a readily applicable technology. There are many examples of attempts to implement PSD in one-off building designs, especially for housing. Some of these attempts have been unsuccessful for a number of reasons. For example focusing too exclusively on solar aspects has compromised performance in a number of instances. This has been particularly true for conservatories built on to houses which have often been incorrectly seen as an easy passive solar option. Other attempts have failed because the solar design principles have been misunderstood or misapplied in the absence of proper design guidance. In some cases the design itself has been too ambitious, incorporating complex or inappropriate measures.

There is very little experience, so far, of successful PSD in mass market housing schemes. There are technical and market related reasons for this, but achieving an impact in this area is essential if PSD is to have a real effect on the domestic energy scene. We still need more detailed knowledge of and design guidance on the more complex PSD issues, particularly for non-domestic buildings as there are fewer examples of PSD in these building types. The barriers to its uptake here are both technical and market-based. A technical understanding of PSD in non-domestic buildings is complicated by the interaction of many design factors. Although the apparent readiness of some designers and their clients to incorporate solar features such as atria suggests a market pull for PSD, designers generally are not yet fully aware of the solar and energy consequences of such measures.

The industry is largely content to carry on with conventional design procedures. This is mainly due to a lack of skills and knowledge of new measures, a cautious approach to new procedures which are perceived as risky, and lack of client motivation.

The Government's Programme

The passive solar programme has been assessing various passive solar design measures to identify, develop and commercialise those which could be cost-effective. The objectives for the programme, embodied in Energy Paper 55, were:

- to evaluate those aspects of the technology that could make significant and economic changes to the energy used in buildings;

- to develop and test design guidance so as to reduce the risks of applying the technologies;

- to produce information on the performance of passive solar design in practice;

- to put together messages about passive solar that will influence building designers, developers and users;

- to develop technology transfer tools to ensure that these messages are understood and taken up effectively.

The main elements of the programme as set out in Energy Paper 55 were:

- a major programme of field trials in housing and non-domestic buildings (the EPA projects);

- an expanded programme of design studies in both the housing and non-domestic sectors;

- a programme of information dissemination;

- a continued programme of underlying R & D and assessment studies;

- an evaluation of the concept of a design advice scheme.

Most of the objectives for these activities have now been achieved. Major achievements are listed below.

- Through the EPA projects, a cost-effective route for obtaining data on real passive solar buildings has been evolved, tested and put into practice. Detailed technical and summary reports have been produced on over 30 buildings.

- The housing design studies have produced over 50 designs of houses using passive solar features along with other appropriate energy efficiency measures. Many of the major housing developers have been involved in the programme as clients or quasi-clients for the designs. The results from these studies provide a comprehensive evaluation of the potential for passive solar in marketable house designs in a wide range of UK locations.

- The housing design guide has been completed and published.

- The non-domestic direct gain design study project has produced nine designs involving direct gain passive solar features, with an emphasis on daylighting and natural ventilation. Each study has involved a commercial design team and client.

- Development work has been carried out in a number of areas relating to daylight design including daylight availability measurements, collation of daylighting calculation algorithms and the incorporation of daylight design considerations into proposed revisions of building regulations.

- Results from the programme are being fed into the Energy Efficiency Office's Best Practice programme for energy efficiency measures in buildings.

- The design advice scheme concept has been tested and evaluated. The pilot scheme in Scotland carried out over 270 consultations with design teams and their clients and it is estimated that the annual energy savings accruing from this are in excess of £700,000. The scheme has now been re-launched as a national Energy Design Advice Scheme with a target of over 2,000 consultations over five years and resulting annual energy savings of £50 million by the year 2000.

The Department of Energy, through the Building Research Establishment (BRE), carries out research across the whole range of building and construction activities. The BRE is an important source of expertise for the Passive Solar programme and acts as a technical consultant and advisor to the programme. The Energy Efficiency Office programme on energy efficiency in buildings is operated by the Building Research Energy Conservation Support Unit (BRECSU) at the BRE. The EEO's Best Practice programme has been established to provide information on good practice in energy efficiency in the buildings sector, including consumption guides, case studies etc. The Best Practice programme has established its credibility in the market as a provider of authoritative and independent information. To take advantage of this, the results of the Passive Solar programme, once developed into robust design guidance, are being fed down the Best Practice route as part of a drive towards an integrated approach to low energy use in buildings. The Energy Design Advice Scheme is managed by BRECSU and through them close liaison is maintained with the activities of the Best Practice programme.

The BRE is carrying out an extensive programme of daylight research for the DoE (Construction Directorate). The work includes the production of a design guide on innovative daylighting systems, measurement of daylight availability, integration of daylight into energy modelling methods, contributions to codes and standards, and the use of daylighting with computer display screens (VDUs).

BRECSU, on behalf of the EEO, is developing the LT (Lighting and Thermal) Method. This is a design tool for use at the sketch stage of a building design which enables an energy optimisation to be made of strategic design decisions. It helps designers to manipulate the building's form and facade design to achieve a balance between lighting and thermal design requirements.

The Department for Education has carried out a small programme of work to develop guidelines for passive solar and other energy efficiency measures in the design and operation of schools. This work has established links with the Passive Solar Programme.

Figure 3: Looe school in Cornwall which uses passive heating and lighting measures.

In addition to these activities, the SERC supports more fundamental building science research activities, some of which provide additional information on the operation and performance of passive solar measures.

The UK has participated in a number of International Energy Agency activities on passive solar design. It has made a major contribution to Task XI of the Solar Heating and Cooling Agreement on Solar Commercial Buildings. This work has been carried out as part of the field trials activities outlined above. Contributions have also been made to the Building and Community Systems Agreement Annex on Building Environmental Performance.

In the European Community, the CEC DGXII and DGXVII programmes have included projects on PSD in which UK organisations have participated. These include the Building 2000 project, the Working in the City competition and the Project Monitor series of monitored case study buildings, as well as co-ordinated programmes of research on particular topics. The network of PASSYS test cells is an important initiative of DGXII. There are now over 20 cells established in 10 countries - including four in the UK – carrying out component and computer model validation measurements.

The CEC is also funding research on daylight measurement, integration of daylight into energy modelling methods and the development of holographic glazing at various European centres (including BRE) as part of the JOULE programme.

Figure 4: Conservatory measurements on a PASSYS test cell.

B.Other Countries

Passive solar design is an active area of R,D&D in many countries. The focus of attention varies according to the main opportunities for application, – heating in northern and temperate climates such as the Scandinavian countries and cooling and ventilation in hot climates such as those bordering the Mediterranean. In most countries, activities centre around the dissemination and promotion of proven measures together with R&D on technological innovation such as advanced glazing. This parallels the UK programme activities outlined above. The largest programme on PSD was mounted in the USA during the late 1970s and early 1980s. It was comprehensive and covered residential and non-domestic buildings across the range of North American climates. Much of the current understanding of solar heating and daylighting stems from the results of this programme, particularly from the work performed by the Solar Energy Research Institute (now National Renewable Energy Laboratory) and the Lawrence Berkeley Laboratory. A Passive Solar Industries Council has been established in the USA to promote PSD throughout the building industry and a number of builders' guidelines have been produced to support this activity. The UK programme has collaborated with the US work under a Bilateral Agreement, the emphasis being on building monitoring, mathematical modelling and data analysis.

Daylight research and demonstration is also underway in other European centres, in Japan, Australia and particularly in the United States. The CIE (Commission Internationale D'Eclairage) and the IEA co-ordinate some of these activities under collaborative agreements.

Geothermal Hot Dry Rock

I. Technology Review

A. Technology Status

There is a large amount of heat just below the earth's surface – much of it stored in low permeability rocks, such as granite. This source of geothermal heat is called 'hot dry rock' (HDR). Once the heat is extracted it would take several thousand years for it to be replenished and therefore on a human time scale HDR is not strictly a renewable energy source.

The concept of HDR technology is based on drilling two holes from the surface then pumping water down one of the boreholes, circulating it through the naturally occurring, but artificially dilated, fissures present in the hot rock and finally returning it to the surface via the second borehole. The superheated water or steam reaching the surface can be used to generate electricity or to supply combined heat and power systems. The two boreholes (a 'doublet') are separated by several hundred metres in order to extract the heat over a sizeable underground volume of rock, termed a 'reservoir', between the two holes, as shown in figure 1.

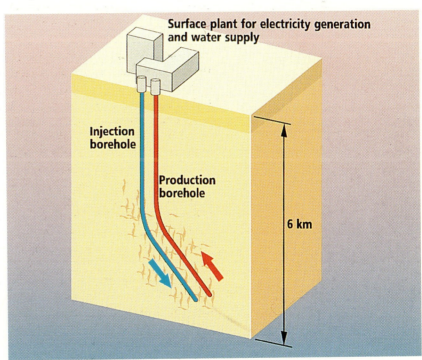

Figure 1: A conceptual geothermal hot dry rock scheme.

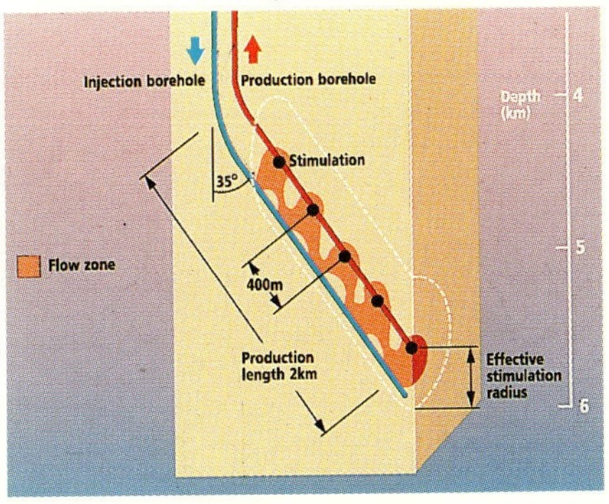

It was initially assumed, following work at Los Alamos in the USA, that a single artificial planar fracture could be created in the low permeability rock by hydrofracturing. Now, however, the consensus is that hydraulic injection needs to be used to reactivate and enlarge pre-existing fractures (or joints). The concept, with multiple reservoir cells, is shown schematically in figure 2.

The engineering of the reservoir has proved to be a formidable technical problem which has not yet been satisfactorily solved after more than ten years of intensive research in this country, the USA and elsewhere. Because of the technical difficulties, there are no commercial HDR

Figure 2: An underground reservoir.

schemes in existence anywhere in the world. If the technology could be successfully developed, a small HDR power station with one doublet might be expected to produce about 4 MW (gross electrical output), on a 10 ha site containing the well heads, control and generation buildings, tanks and lagoons. A full-scale commercial power station might comprise a number of operating reservoirs, producing a total of about 50 MW of electricity and be expected to operate for at least 20 years. Once drilling has been completed, visual intrusion would be relatively small.

B. Market Status

UK contractors are among the world's most technically advanced in the HDR field and have participated in R&D projects in many other countries. However, after more than 15 years of effort, the research has yet to progress to an operational HDR prototype system. The technology must therefore be considered as speculative.

Current Status	Research
UK installed capacity	Nil
World installed capacity	Nil
UK industry	Several specialist companies

Table 1: Status of the technology.

C. Resource

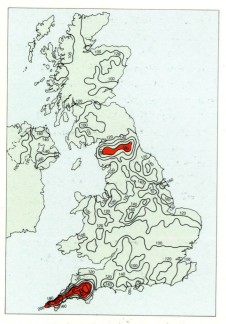

Figure 3: Estimated subsurface temperatures throughout the UK at a depth of 6 km.

The British Geological Survey has predicted the temperatures at various depths throughout the UK. Their results for a target depth of 6 kilometres indicate that the highest temperatures are located in the granite regions of southwest England and Weardale (North Pennines), as shown in figure 3. Initial optimism of a widespread resource proved to be unfounded, as temperatures of 200°C, necessary for electricity generation, are not accessible with current drilling techniques and are not likely to be available in the medium term. A comprehensive heat model exists for southwest England but in the absence of detailed drilling data and geophysical data, a uniform geothermal gradient has had to be assumed for Weardale.

The heat in the rock can be converted to output electrical power at an overall efficiency of only 3%. Using the HDR Cost Model developed by Sunderland University, which simulates a mature, widely applicable technology, it is possible to estimate the distribution of the resource in terms of power output per sq.km.

The UK Accessible HDR Resource that could be exploited by a mature technology, omitting nationally-designated landscape areas (e.g. national parks) and regardless of cost, is estimated to be:

	Accessible Resource (TWhe)
South West England	1030
Weardale	470
Total	1500

If this resource were to be exploited over 25 years it would result in 60 TWhe/year or 7,600 MW of net output power at 90% availability. The distribution of the Accessible Resource in south-west England is shown

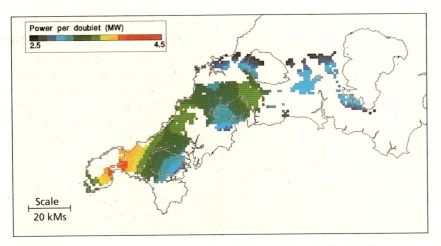

Figure 4: Location and power output of on-land Accessible Sites in SW England with >2.5MW power output per doublet.

in figure 4. At 1991 prices, it has been estimated that the cost of generating electricity in the long term could be about 17p/kWh at 8% discount rate assuming HDR systems perform to conceptual design specifications and theoretical operational conditions. For electricity generation at under 10 p/kWh (1991 prices) and a discount rate of 8% estimates suggest that there is no Accessible Resource for HDR in the UK.

D. Environmental Aspects

Water Supply

The main impact an HDR electricity generating station might have on a local community arises from the large water supply required. No significant problems would be expected from the water losses incurred during stimulation and circulation of an HDR system. However, the initial provision and replenishment of stimulation volumes, and the long-term supply of routine make-up water, would require major raw water storage reserves on site. Local rivers in Cornwall have insufficient flow rates to support the water requirements of a major programme of HDR geothermal installations and any such development would ultimately have to depend on supplies of sea water. The issues involved have not so far been studied in any detail, but the use of sea water in HDR systems appears to be technically feasible, but would carry a cost penalty.

Land Impact

The surface area for each HDR doublet is likely to be 1 sq.km of which the power station would occupy 10 ha. Should it be necessary to use sea water for cooling, then the laying of pipelines would cause some disruption.

Visual Intrusion

The above-ground appearance of an HDR site would include drilling rigs (during the construction phase) and power generation buildings. It would probably be necessary to avoid national parks, though there may be opportunities to site the power stations non-intrusively, for example in disused quarries.

Emissions and Pollution

Although no environmental problems have arisen from the use of gels in stimulations at the experimental stage, the possible use of gels in commercial operations would require detailed consideration by the regulatory authorities. The potential for atmospheric pollution is small, though the organic fluids currently proposed for use in the power plant have given problems in the past when accidentally released to the atmosphere.

Noise

Some noise would be produced by pumps, turbines and generators housed in a building, but noise from drilling would be the major impact. For a doublet, this would be produced over a period of about eighteen months during the construction phase.

Environmental burden	Specification
Gaseous emissions:	
Carbon dioxide	None
Methane	None
Nitrous oxide	None
Chlorofluorocarbons	None
Sulphur dioxide	None
Nitrogen oxides	None
Volatile organic compounds	None during normal operation, but could have accidental release from power plant
Ionising radiation	None
Wastes:	
Particulates	None
Heavy metals	None
Solid waste	None
Liquid waste	None
Thermal emissions	45 MW from evaporative cooling tower
Amenity & Comfort:	
Noise	Low
Visual intrusion	None except during drilling
Use of wilderness areas	0 to low
Other	Disruption from laying of pipelines from coast if seawater is used

Table 2: Environmental burdens associated with Geothermal HDR.

In a speculative 50 MW commercial power station, each component doublet would produce about 4 MW gross, 3.3 MW net electrical power. At 90% availability this is equivalent to 26 GWh/year per doublet. The emission savings, assuming that each kWh generated by HDR displaces a kWh generated by the average plant mix in the UK generation system (for the year 1990) and thereby saves its associated emissions, are summarised in table 3.

Emission	Displaced Emissions (tonnes of oxide/GWh) scheme (tonnes of oxide)	Annual savings per typical 50 MW net electrical power
Carbon dioxide	734	290,000
Sulphur dioxide	10	3,950
Nitrogen oxides	3.4	1,340

Table 3: Emission savings relative to generation from the average UK plant mix in 1990.

E. Economics

The cost of power generated from HDR has been the subject of several major studies. The most recent are a conceptual study for an HDR prototype system by RTZ Consultants in 1990 and the development of the HDR Cost model by Sunderland University from 1988 to 1992, to simulate the performance of the

Season	Day/Night
Spring/Autumn	90%
Summer	90%
Winter	90%

Table 4: Load factor.

technology when developed to maturity. The power plant used for the purpose of this cost analysis is based on a post-prototype HDR system, using an Ormat binary plant to generate power. This represents a compromise in that assumptions have been made about reservoir performance which have no basis in experience; these have then been applied to a system coupled to developed power generation technology.

At current prices, the estimated cost of power generated by a system based on mature HDR technology is several times more than conventional sources.

F. Resource-Cost Curves

Even assuming a technical breakthrough in reservoir creation, institutional and environmental constraints mean that public perception and planning consents would limit the rate at which exploitation could proceed. Another vital factor is the availability of a year-round, plentiful water supply. These factors indicate that a more realistic estimate of the maximum rate of exploitation, expressed as the Maximum Practicable Resource, is 4 TWh/year, or 500 MW, and this is mainly located in Cornwall. With a mature, working technology, this rate of heat extraction could be maintained for several hundred years. A resource-cost curve has been derived for the Maximum Practicable Resource. Given these estimates of the cost of HDR-generated electricity, no uptake is currently foreseen, even under the most optimistic forecasts.

In 2005 the technology will still be at commercial prototype stage, so no contribution will be available. In 2025 it is assumed that if R&D yields an effective HDR technology, three power stations could be in operation, totalling 150 MW capacity at 90% availability. It is assumed that the most geothermally favourable sites would be exploited first, in which case the minimum cost will be 14p/kWh at 8% discount rate and 20.3p/kWh at 15%, as shown in the resource-cost curve. However, if development were to proceed in locations with "average" geothermal gradients within the identified area of potential resource, the cost could be taken to be the mean of a possible wide range, i.e. 22.6p/kWh at 8% discount rate and 40.1p/kWh at 15%.

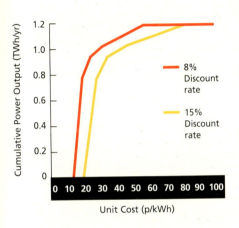

Figure 5: UK HDR Maximum Practicable Resource – 2025

G. Constraints and Opportunities

Technological / System

A conceptual design study for the construction of a deep HDR prototype and associated power generation station in Cornwall was carried out by RTZC in 1989 to 1990. The work was intended to validate the feasibility of engineering an HDR heat exchanger suitable for commercial operation and to define the programme plan for the development of a deep HDR prototype. The main conclusions of the study are the following.

- The geological environment at 6 km depth, particularly the characteristics of existing joints and fractures, is highly conjectural and will not be known until drilling to these depths is carried out.

- If technical viability is to be established, it can only be achieved through an extended research programme relevant to a deep HDR system. This will require exploration to, and experimental work at, 6km depth.

- Even if the viability of the technology were demonstrated, each and every HDR system would, due to geological variations, carry significant risk of technical failure as distinct from the certain repeatability of successive units of a conventional generation programme.

- The drilling performance specification is extremely arduous and outside the experience, and possibly the capability, of current drilling companies. However, further developments in drilling technology could lead to improved performance and cost reductions.

- Further work is needed on the industrial development of well completion techniques, stimulation gels and short-circuit sealants reliable under the conditions expected at 6 km. While these developments may produce some cost reductions, sensitivity studies using the HDR Cost Model suggest that there is little scope for cost savings.

- The site-specificity of the technology reduces the opportunities that can arise from the routine application of generic techniques. Furthermore, it is believed that rock mass characteristics can be determined only by drilling to the target depth. This would place a heavy burden on those schemes where the boreholes prove to be unsuitable.

Institutional

The immaturity of the technology and the significant uncertainties make HDR a high-risk investment. RTZC concluded that any expectation of private capital investment in the commercial development of the technology is unrealistic in the foreseeable future.

The expertise developed as a result of the UK HDR programme is state-of-the-art and as such may be attractive to foreign projects. Whether significant financial benefit can accrue from such international transfer of technology is uncertain.

H. Prospects Modelling Results

Due to its current state of technical development, poor economics and the small size of the resource, Geothermal HDR is not expected to make any contribution to the energy supplies between now and 2005. Beyond this time period its potential is uncertain, but it has been speculated that the technology may be developed sufficiently by 2010 to start building power stations at the rate of one or two 50 MW stations every five years, up to a maximum of ten operating simultaneously. It is very unlikely, however, that these plants would be economically viable, even under the most favourable scenario for renewables.

II. Programme Review

A. United Kingdom

The UK HDR Programme has established the location and likely size of the resource and has confirmed that electricity generation is the most cost-effective option. Experimentation has been concentrated at Rosemanowes, Cornwall, shown in figure 6.

Work started in 1977 with a demonstration of the feasibility of establishing subsurface hydraulic connections between shallow (300 m deep) boreholes drilled into the granite. This was sufficiently successful to justify the drilling of three boreholes to a depth of between 2 and 2.5 km. From the evidence of extensive circulation experiments, the reservoirs created between these wells proved to have unacceptable performance, in terms of accessible volume, thermal drawdown, and impedance.

The results of a review in 1990 indicated that the option of going ahead with a programme to construct a 6km HDR prototype in Cornwall was no longer a feasible proposition. In light of the review's conclusions, the Government decided that a new direction of the programme was necessary. The current phase of work, from 1991 to 1994, has concentrated less on research in Cornwall and involves greater collaboration with a European programme already under way with France, Germany, and the European Commission.

However, a practical method of extracting useful energy from HDR is still a long way in the future. So far no country in the world has created an

Figure 6: The Rosemanowes Project.

underground HDR reservoir with operational characteristics which give any confidence that a commercial-scale system could be successfully created.

The cost of power generated from HDR has been assessed with the use of the conceptual study for an HDR prototype system produced by RTZ Consultants, and the development of the HDR Cost Model by Sunderland University from 1988 to 1992. The heat in the rock can be converted to electrical power output at an overall efficiency of only 3%. At 1991 prices, it has been estimated that the cost of generating electricity could be about 17p/kWh at an 8% discount rate. Using the HDR Cost Model, which simulates a mature, widely applicable technology, it is possible to estimate the distribution of the resource in terms of power output per sq.km. The review has also considered institutional and environmental constraints and concluded the rate at which exploitation could proceed would be further limited by these factors.

B. Other Countries

The UK is participating in the European HDR Programme, with all downhole work taking place at Soultz near Strasbourg in France, and at the German spa town of Bad Urach. HDR contractors funded by the UK, France, Germany and the European Commission have formed the European Hot Dry Rock Association (EHDRA) in order to adopt a co-ordinated approach to R&D. The principle activities at each site are as follows:

- Rosemanowes   instrumentation development, numerical modelling, geophysical techniques;

- Soultz   deepen well GPK1 from 2 km to 3.5 km, characterise rock mass, stress-measurements and assess flow potential;

- Urach   deepen well from 3.5 km to 4.4 km, characterise rock mass, stress-measurements.

The results of the tests will be incorporated into feasibility studies. Work at Urach has been suspended following problems during extension of the well. Plans for further experimental work are therefore being reconsidered. With a similar report on Rosemanowes, these feasibility studies will be compared, leading to selection of the site for reservoir creation studies and, in 1994/95, a recommendation of the site for construction of a European HDR prototype.

Major projects are underway in the USA and Japan. At Fenton Hill New Mexico, the Los Alamos HDR project initiated a long-term flow test in 1992, circulating water through the reservoir created in 1984. The purpose of the test was to demonstrate that heat could be extracted from the system over an extended period. The target was two years, but progress was interrupted by technical problems. Further progress in the USA depends on the resumption of the long-term test. A decision on whether or not to construct a commercial prototype will depend critically on the results of that test.

In Japan, the results of two projects have been sufficiently encouraging to justify proceeding with further drilling and reservoir creation experiments. At depths of less than 2 km, rock temperatures exceeding 200°C have been measured. Reservoirs have been created, but only between wells separated by a short distance. To achieve flow and temperature targets comparable with commercial systems, wells are being deepened and further stimulations are planned. The project at Yunomori represents a diversion from true HDR since the rocks contain hot fluids but these require stimulation through water injection to extract economic amounts of heat. This so-called 'Hot Wet Rock' project is scheduled to begin circulation tests next year.

Geothermal Aquifers

I. Technology Review

A. Technology Status

Geothermal aquifers exploit heat from the earth's crust through naturally occurring ground waters in deep porous rocks. The exploitation of these aquifers as a source of energy requires a production borehole to extract the water and an injection hole to dispose of the cooled water. An alternative single hole configuration can be used where instead of using an injection well the used water is simply discharged to the sea or some other convenient sink. Because of the poor thermal conductivity of rock and low-fluid recharge rates, heat is usually extracted at a greater rate than it is replenished from the surrounding rock mass. Geothermal aquifers are, therefore, not 'renewable' resources in the strict sense of the word, but are usually grouped along with renewables.

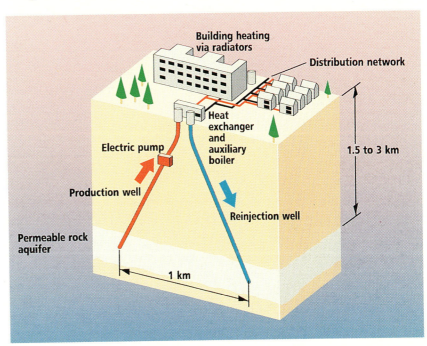

Figure 1: A typical geothermal aquifer scheme.

Waters at moderate temperatures, in the range 50°C to 150°C, can be used for heating purposes, for example in a district heating scheme, as shown in figure 1. In France, where geothermal district heating is widely established, over sixty schemes are in operation, with an average geothermal contribution of 57 GWh_{th}/year, serving some 200,000 dwellings. Elsewhere (such as in the USA, Philippines and Italy) higher temperatures allow electricity to be produced. A typical plant size is around 55 MW_e and often several of these are combined in a single generating station.

B. Market Status

The use of aquifers to produce electricity or to provide heat is well established in those areas of the world with suitable geological conditions. These usually occur close to the crustal plate margins. About 6,000 MW of electricity generating capacity is currently installed around the world in places where both steam and water are produced at temperatures over 200°C. Half of this capacity is located in the USA. World-wide some 11,000 MW of geothermal energy is used for district or process heating.

Water temperatures in the UK are too low for electricity generation. The only operating geothermal scheme in the UK, an exploratory well funded by the UK Government and the CEC, is at Southampton, where the geothermal aquifer contributes 2 MW of heat into a 12 MW district heating scheme in the centre of the city.

Current Status World-wide: deployed	UK demonstration
UK installed capacity	2 MW$_{th}$
World installed capacity	6,000 MW$_e$ 11,000 MW$_{th}$
UK industry	Small equipment or drilling market for large companies involved in other areas

Table 1: Status of technology.

C. Resource

Figure 2: Resource map of Upper Palaeozoic and Mesozoic Basins.

In the UK, very little of the available water is above 60°C and so the resource is not suitable for electricity generation. In principle, it could be used for district heating with top-up and standby heating from coal fired boilers or to substitute for local boilers using gas, coal or oil. The potential in the UK is limited both by the low water temperature and the availability of suitable heat loads adjacent to the water reservoirs. The resource that does exist is found mainly in Permo-Triassic sandstone basins, as shown in figure 2.

The potential geothermal fields identified by the British Geological Survey are shown in figure 3.

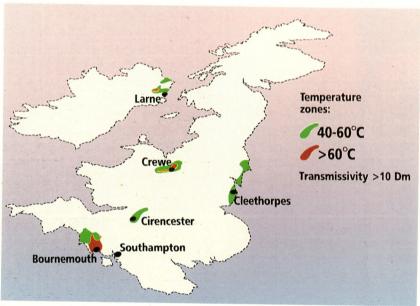

Figure 3: Resource map of potential geothermal fields in the UK.

Exploitation is limited to areas where the heat loads coincide with the resource, as it is not economically feasible to transport the hot fluids any significant distance. Based on a detailed assessment of the heat requirements of Grimsby, and applying the results proportionally to the other potential geothermal fields, it is unreasonable to anticipate more than a total of 50 geothermal schemes being developed in the UK. Assuming that each viable scheme would derive an annual minimum of 26 GWh$_{th}$ from a geothermal source, the Accessible Resource is 1,300 GWh$_{th}$/year.

However, the geological conditions at the depths of interest are not known in detail and there would be considerable uncertainty about the lifetime of any proposed scheme until pumping tests had been carried out. The aquifer may be physically constrained by boundaries such as faults or less permeable strata of rock. This would cause the flow of hot water to cease after a shorter time than expected, or the temperature of the water to be below that anticipated.

Discount Rate	England & Wales	Scotland	Northern Ireland	UK
8% and 15%	1.05	0	0.25	1.3

Table 2: Accessible Resource (TWh$_{th}$/year) in the UK at less than 10p/kWh$_{th}$ (1992).

D. Environmental Aspects

For any district heating application the drilling operations would have to be conducted near to or within cities. The noise from such operations is a potential problem but probably not any more severe than that encountered with other major construction projects. Considerable local disruption would occur if district heating mains needed to be laid in existing urban areas. Once a scheme is installed there should be no disturbance except during repairs or maintenance operations.

The water discharged from the heating scheme may contain significant concentrations of dissolved salts. In disposing of these, a second borehole would be essential, as care must be taken to ensure that they do not contaminate other boreholes used for public water supply. The majority of potential UK schemes would therefore be doublets; the exceptions would be at coastal sites where singlets, with water discharge to the sea, could be used.

Environmental burden	Specification	Environmental burden	Specification
Gaseous emissions:		Wastes:	
Carbon dioxide	None	Particulates	None
Methane	None	Heavy metals	None
Nitrous oxide	None	Solid waste	None
Chlorofluorocarbons	None	Liquid waste	None[1]
Sulphur dioxide	None		
Nitrogen oxides	None		
Carbon monoxide	None		
Volatile organic compounds	None		
Ionising radiation	None	Thermal emissions	None
Amenity & Comfort:			
Noise	None[2]		
Visual intrusion	0 to low		
Use of wilderness areas	0 to none		
Other	None[3]		

Table 3: Environmental burdens associated with geothermal aquifers.
Notes: 1: Zero if re-injected, or 10 to 50 litres per second of saline fluids if discharged.
2: Except during drilling.
3: Disruption due to laying heat mains.

Emission	Displaced Emissions per GWh (tonnes of oxide)	Annual savings per typical 2 MW$_{th}$ scheme (tonnes of oxide)
Carbon dioxide	370	6,000
Sulphur dioxide	2	32
Nitrogen oxides	1	16

Table 4: Emission savings. These figures assume geothermal aquifers displace heat which would otherwise have been provided by a domestic gas boiler. The emission savings would be greater if electrical heating were to be displaced.

E. Economics

Geothermal heating schemes are capital intensive; capital and non-geothermal operating costs account for 75% of the total scheme cost. The cost of a single borehole represents only about 20% of the total cost.

Geothermal aquifers having both a production and re-injection well (known as a doublet) are unlikely to be cost-effective in any scenario except for the highest energy price considered.

F. Resource-Cost Curves

To estimate the uptake of the potential resource, it is necessary to make an assessment of the number of schemes likely to be developed. This would depend on the experiences of existing schemes (i.e. Southampton) and the possibility of support (for example, from the CEC Thermie programme). A subjective assessment suggests that if one scheme of 26 GWh_{th}/year were developed in each potential geothermal field, then five schemes may be completed before 2005, which could then be followed by development of a further 20 schemes by 2015. The Maximum Practicable Resource is then:

Season	Day/Night
Spring/Autumn	75%
Summer	66%
Winter	100%

Table 5: Load factor.
Note: the load factor has been used to adjust the output of scheme (pumping rate in summer is 66% of that in winter) and the heating season in the UK has been estimated at 240 days.

130 GWh_{th}/year by 2005   650 GWh_{th}/year by 2025

The cost of exploitation has been estimated assuming that notional schemes with similar heat production rates are developed in each area of potential resource. Exploitation would depend on the specific characteristics of the aquifers important factors such as fluid supply, rejection temperatures and pumping rate may vary greatly. It can be assumed that environmental standards will dictate that in all new schemes discharge of cooled fluid will be by injection to the source aquifer.

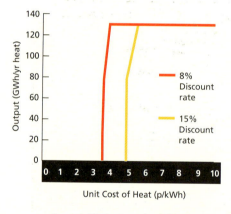

Figure 4: UK Geothermal Aquifer Maximum Practicable heat resource – year 2005 based on doublets.

Each scheme would possess a geothermal installed capacity of 10 MW (50% of the peak demand of the heating scheme). A cost element has been included for surface plant, allowing for the installation of heat mains, heat stations including heat pumps, control systems, etc. Operating costs include management and maintenance, with power consumption dictated by the role of the heat pump in the scheme. Minimum cost of heat is estimated to be 3.5p/kWh_{th}. This cost assumes that viable wells can be drilled into the aquifer without failure – no allowance has been made for the real risk of unusable wells being drilled.

The cost of heat from aquifers can be put into context by comparison with the unit cost of heat from conventional fossil fuelled industrial boilers of approximately £4/GJ (i.e. 1.44 p/kWh_{th}) which offer considerably less technical risk. In practice, there would be a high risk in developing a geothermal aquifer in the UK, largely because transmissivities and heat flow are unlikely to be uniform. Moreover, it would not be possible to accurately predict these characteristics in advance of downhole tests, which would also be required to determine whether heat-flow could be sustained over the lifetime of the project.

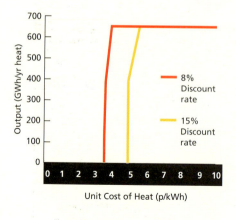

Figure 5: UK Geothermal Aquifer Maximum Practicable heat resource – year 2025 based on doublets.

Commercial development of a system with high technical risks is likely to demand higher rates of return than would be expected for conventional systems. Furthermore, any investment plan would have to include costs for major technical contingencies, if for example complications arose during drilling. These factors are not reflected in the Maximum Practicable Resource/cost curve (figure 4). Given the disparity in the risk and cost of heat from aquifers and conventional industrial boilers, the former are unlikely to be commercially developed in the foreseeable future.

G. Constraints and Opportunities

Technological / System

There are no major technological constraints to the development of geothermal aquifers since the technology is proven. In the UK the major constraint is the size and quality of the resource. In other countries (e.g. USA, Japan, Italy) where the resource is quite extensive and of a higher quality than in the UK, geothermal aquifers provide a significant contribution to energy requirements.

The technical risk with such schemes remains high, due to the uncertainty surrounding the life and performance of any potentially exploitable aquifer.

Institutional

The market potential for aquifer-based heating systems is limited by the lack of an established district heating infrastructure. However there could be more scope in other applications such as horticulture and fish farming which do not require such distribution networks.

H. Prospects Modelling Results

Under current estimates of future energy requirements and costs of competing sources of heat, no further uptake of heat from geothermal aquifers takes place beyond the existing contribution of 16 GWh_{th}/year.

II. Programme Review

A. United Kingdom

Commercial interest in geothermal aquifers is very low in the UK at the present time. The district heating scheme in Southampton, shown in figures 6 and 7, incorporates the sole example of an operating scheme.

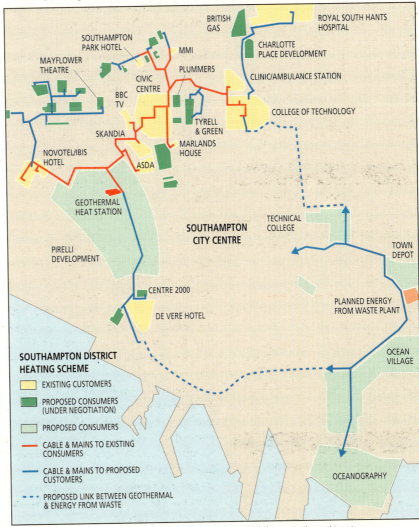

Figure 6: Map of the Southampton district heating scheme which has a geothermal input.

Figure 7: Southampton's district heating scheme's heat station. (*Courtesy of Paul Carter Photography.*)

The principal barrier to commercialisation is the limited and uneconomic aquifer resource in the UK. In addition, the risks associated with exploration are relatively high. This was demonstrated by the disappointing results from the former Department of Energy's exploration programme carried out in the late 1970s/early 1980s. The techniques for extracting heat from aquifers are well established in those countries which have a significant resource, so the technology is proven overseas.

B. Other Countries

Overseas R&D concentrates on two major issues: the behaviour and performance of currently exploited reservoirs and the analyses of the geochemical factors in production and re-injection wells. The declining performance of some geothermal fields, especially the Geysers in California, has highlighted the fact that the resource is finite and that overall co-ordination of exploitation is necessary to minimise the possibility of over-extraction.

Municipal and General Industrial Wastes

I. Technology Review

A. Status of the Technology

Wastes produced by households, industry and commerce pose a difficult, costly and potentially environmentally damaging problem. Using the materials to produce energy can reduce these environmental problems, and at the same time save fossil fuels. In order to be cost-effective the technology needs to provide energy at a cost competitive with other sources, but also provide a disposal route that can compete on cost with other available options.

The energy content can be recovered by a number of technologies, for example:

- by combustion of the raw waste (incineration);

- by combustion after processing to remove undesirable components or materials that can be recycled (refuse-derived fuel);

- by biological processes that occur spontaneously when the waste is landfilled (landfill gas collection and utilisation);

- by biological processes under controlled conditions in digester systems (anaerobic digestion).

The potential for recovering energy from these wastes, and the mix of the technologies that will be employed in the future, is inevitably bound up with waste disposal practice.

Figure 1: Waste processing at a modern sorting facility.

Exploitation has been through successive cycles of development and decline. Systematic incineration of household waste, for example, was pioneered in the UK in the late 19th Century, initially as a means of sanitary disposal in urban areas but subsequently also as an opportune means of electricity production. Indeed, by 1912 some 76 plants in the UK generated power from waste in this way. Similarly, methane produced from the digestion of sewage sludge has long been used to power water treatment works.

The current situation in the UK is that over 90% of the waste produced by households and commerce is disposed of to low-cost rural landfill. Regulatory pressures are set to tighten landfill standards and drive up costs, whilst recycling and waste reduction initiatives will reduce the quantities of waste requiring disposal. While such pressures will influence trends in waste disposal, the continued availability of suitable landfill sites in the UK will ensure that landfilling will remain the major route for UK waste disposal. The exploitation of landfill gas will thus remain a major option for the recovery of energy from wastes. This is dealt with in the Landfill Gas module.

Incineration

Incineration is an established means of processing wastes. The principal aim is to reduce the waste to a small volume of sterile ash and thereby provide savings in transport costs and landfill requirements. This is widely practised in many parts of Europe. The energy is recovered and used for power production, or for combined heat and power (CHP), with the heat often used in district heating schemes. The technology is well established and fully commercial, but the capital and operating costs of the plant are high. In many European countries landfill is not favoured and the higher costs of treating wastes by incineration prior to disposal are accepted as inevitable. Such plants typically handle between 35 and 80 tonnes of waste per hour (250,000 to 600,000 tonnes per year) and produce between 15 and 40 MW_e.

Small-Scale Incineration

Commerce and industry also produce general industrial waste along with production wastes and other, more specialised, wastes. These wastes tend to have a higher calorific value than household waste, and there are greater opportunities to segregate uncontaminated or good quality materials. Such wastes are principally handled by the private waste disposal industry and there may be opportunities for their use in smaller local incineration plant to produce electricity. A typical plant would use 5 to 10 tonnes of waste per hour and produce 5 to 10 MW_e. The technologies for such utilisation are discussed in the Specialised Industrial Wastes module.

Recent legislation will lead to a greater involvement of the private disposal industry in the household waste sector and this will tend to blur the distinction between wastes of different origin. As with overseas practice, this may lead to commercial and industrial wastes entering the household waste stream for treatment in larger-scale facilities. This module therefore includes the assessment of this potential.

Alternative Technologies

A range of alternative technologies for utilising wastes are under development. These may compete on cost with landfill, be at a scale appropriate to local conditions and/or provide environmental advantages over the more mature technologies. In addition these new technologies could be more compatible with the move to increased levels of waste recycling, whether via separate collection of various waste components or via the separation of recycled materials at central facilities.

Figure 2: Densified refuse-derived fuel (d-RDF) production.

Techniques for producing and using fuels generated from refuse, refuse-derived fuels (RDF), have been under development for some years. A wide range of possible flow sheets have been proposed, ranging from schemes in which waste is pulverised and used with little additional treatment (coarse RDF), through to schemes in which the waste is extensively processed to allow recovery of recycled materials and production of a refined, dried and densified fuel product (densified RDF). Developments in the UK have largely focused on the production of d-RDF aimed at the industrial boiler market. Several full-scale fuel production plants have been built and operated, but most have experienced difficulties in building up markets for the fuel. Interest is now focusing on processing waste to produce a fuel for electricity production. Such a process would not need to involve the expensive drying and densification stages involved in d-RDF production. Such plant would tend to be modular and handle between 15 and 45 tonnes of waste per hour (100,000 to 300,000 tonnes per year) and produce between 5 and 15 MW_e.

There is a growing interest in the development of biological processes for recovering energy-from-wastes by anaerobic digestion. The waste would be processed to concentrate the biodegradable fractions, then digested in a tank to produce methane for electricity production or for export to local fuel users. This follows conventional practice in waste water industries for the treatment of sewage sludge. These techniques could be used to process the organic fraction of household and commercial wastes, either as part of a source separation scheme or as an adjunct to centralised plant set up for recycling or to produce RDF.

Figure 3: Anaerobic digestion at a sewage treatment works.

In the longer term it is possible to envisage a plant in which waste processing, combustion and biological treatment are used in a concerted way to make best use of the resources in the waste, at the same time minimising environmental impacts. Such Resource Recovery plant could also encompass waste recycling and reclamation activities.

B. Market Status

The current technical development of options for energy-from-waste owes much to oil price rises coupled with a surge of environmental awareness in the 1970s, followed by a further tightening of environmental standards in several industrialised countries in the late 1980s. Notably:

- interest in engineered landfilling, 'biofilling' and anaerobic digestion stem largely from environmental legislation related to ground-water pollution and uncontrolled methane emissions from conventional landfills;

- incineration and flue gas cleaning technology has advanced considerably in response to stringent air pollution standards;

- gasification, liquefaction and other advanced thermal processing technologies are being developed to achieve lower overall emissions and greater conversion efficiencies.

Development has generally taken place most rapidly in the countries which first imposed the tighter standards upon themselves. This is particularly the case with respect to combustion-based technologies, where Germany and the Scandinavian countries hold the technological lead following extensive development in response to tightening air emissions standards in the 1980s.

The status of the technological options is summarised below.

- Mass burn incineration of municipal and general industrial wastes is well established but, under stringent environmental standards, may only be economically viable where sufficient quantities of wastes are available to support large-scale schemes. European manufacturers are considered to hold the technological lead, with Scandinavian manufacturers also providing advanced gas cleaning systems. Several UK companies are licensees for advanced technologies. In the UK, there are 34 large-scale municipal solid waste incinerator plants, mostly built in the 1960s and

1970s. Only five of these are fitted with energy recovery systems. The emission performance of these plants falls well below the new standards, and many of the plants will close or need extensive refurbishment in the next few years. In some cases there will be an opportunity to include an energy recovery system as part of the refurbishment package.

- Resource recovery, involving the integration of centralised processing for materials recovery with energy production from the process residue, is increasingly being adopted overseas for waste disposal at intermediate scales of operation, typically of the order of 200,000 tonnes/year throughput. The component systems are all available in the UK but need to be assembled and demonstrated in practice.

- The high costs of small-scale combustion technologies may be most constrained economically by the requirements of comprehensive emissions abatement. Opportunistic uses or applications may continue where alternative disposal routes are restricted and/or incur a high cost. In particular there is renewed interest in sewage sludge incineration in the UK following international agreements to eliminate disposal of sludge at sea by 1998, and in clinical waste incineration. There are a number of long-established UK manufacturers, but they face increasing competition from advanced technologies from overseas, particularly from Germany and the USA.

- Thermochemical conversion technologies are at an early stage of commercial development. However, they offer significant potential to exploit biomass and waste fuels with a higher overall conversion efficiency and an overall lower impact on the environment. They are discussed in the Advanced Conversion module.

- Anaerobic digestion is conventional technology for farm and sewage sludge treatment. Several commercial schemes treating the organic fraction of household waste have been developed in Europe. The UK has experience at the pilot stage, but it is a relatively high cost technology.

Figure 4: SELCHIP: South East London Combined Heat and Power waste combustion plant.

	UK status	Installed capacity (MW$_e$)		Status of UK industry
		UK	World	
Mass incineration	Commercial	48	>5,000	UK licensees of foreign technologies
Sewage sludge incineration	Commercial	0	Unknown	UK licensees of foreign technologies
Refuse Derived Fuel d-RDF c-RDF	Commercial Development	4 0	<100 >1,000	Several UK suppliers UK licensees of foreign technologies
MSW digestion	Research	0	<10	Collaboration with overseas suppliers. Some UK research interest
Sewage sludge digestion	Commercial	91.4	Unknown	Well established
Thermochemical processing	Research	0	Some, mainly on biomass	Some UK research interest

Table 1: Status of technology.

NFFO has done much to stimulate commercial interest in energy-from-waste, particularly in MSW incineration which has attracted 14 schemes with a capacity of 302 MW$_e$. However, many of the schemes accepted under the second tranche have faced difficulties in securing planning permission and/or waste supply contracts. Delays in securing these contracts and permits are severely eroding the project economics, and already several schemes have been withdrawn.

C. Resource

The best available estimates indicate that there are some 25 million tonnes of household wastes, 25 million tonnes of combustible commercial and industrial waste and 1.5 million tonnes of sewage sludge (dry basis) generated in the UK each year. Household waste has an energy content of around 9 GJ/tonne and the commercial waste around 16.5 GJ/tonne (cf. coal is 27 GJ/tonne). The total energy content of the collected wastes is therefore around 24 million tonnes of coal equivalent per year (Mtce/year), around 640 PJ/year. The useful energy that could be produced from these wastes would depend on the technologies involved. Table 2 gives the conversion factors for a range of possible technologies and estimates the total energy yield that could be produced if all wastes were handled by that route. In practice of course there will be a mix of technologies in use.

Option	Energy Yield (kWh/tonne)	Total Potential (TWh/year)
Incineration:		
MSW	450 to 500	12
Commercial and Industrial	600 to 900	15 to 22
Sewage sludge (dry)	100 to 300	0.2 to 0.5
d-RDF (/tonne MSW)	285	9.5
c-RDF (/tonne MSW)	500	16.7
Anaerobic digestion:		
MSW (/tonne MSW)	80 to 160	2.7 to 5.3
Sewage sludge (dry)	560	0.8

Table 2: Resource assessment in the UK.

These figures assume that all available waste is converted by the indicated technology. All the resource would be accessible at a cost of 10p/kWh or less at both 8% and 15% discount rates. The total figures are for the UK as a whole; regional contributions will be related to population. The current Accessible Resource, assuming all waste could go to incineration as the most efficient process, is approximately 31.5 TWh/year. This figure will change with time, the view of the future and the mix of plant utilising the resource.

There are likely to be changes in the quantities and composition of waste in the future as pressures to minimise and recycle waste take effect. The proportion of waste being landfilled will also change as the other technologies are used. A discussion of the likely effect of all these changes on the potential and cost of energy from these wastes is presented in Section F. The factors that will influence these trends include:

- moves towards waste minimisation and higher levels of recycling, perhaps involving segregation at source;

- developments in environmental legislation affecting both landfill practice and the other disposal options;

- changes in the commercial structure of the waste disposal industry, and particularly in the balance between the roles of the private and public sector.

There is uncertainty about how these changes will affect the potential for recovering energy from these wastes, but these factors will probably be as important as developments in energy economics and policy.

D. Environmental Aspects

Environmental burden	Specification (units per kWh_e)
Gaseous emissions:	
Carbon dioxide	1.6 kg
Sulphur dioxide	2.8 g[a]
Nitrogen oxides	3.2 g[a]
Carbon monoxide	0.9 g[a]
Volatile organic compounds	0.2 g[a]
Particulates	0.3 g[a]
Heavy metals	0.01 g[a]
Ionising radiation	None
Solid and liquid waste	0.7 kg controlled waste
Thermal emissions	3.5 kWh[b]
Amenity & Comfort:	
Noise	Minimal
Visual intrusion	Some – subject to planning
Use of wilderness area	None
Other	Some – transporting waste to and from sites

Table 3: Environmental burdens associated with waste incineration assuming current trends continue.
Notes: a: New plant standards, UK. The emission standards may be bettered in practice by the application of the latest technology.
b: Thermal emission potentially significant for large plant (some could be recovered as useful heat)

All available waste disposal options have some impact on the environment, but the energy recovery options can generally offer environmental advantages over landfill without gas recovery. It is difficult to compare the effects in any quantitative way, but the major issues associated with each technology are set out above. In addition to these, any centralised waste treatment plant would inevitably have some impact on the local environment, through lorry movements, etc. In many cases these local effects give rise to considerable local opposition to the development of projects. Typical environmental burdens from an MSW incinerator operating to latest environmental standards in the UK are indicated in table 3.

Generating electricity by incinerating municipal wastes displaces emissions from conventional generating plant and also averts emissions which would have occurred had the waste been disposed of in landfills. The calculation of such saving is subject to wide uncertainties. However, for the purposes of this analysis, it is assumed that (i) the content of fossil-fuel derived carbon in the waste is negligible; (ii) incineration takes the place of landfilling where landfill gas is collected and flared; and (iii) methane escaping from landfills has a global warming potential eleven times greater than carbon dioxide (as the direct global warming effect over 100 years, IPCC 1990 and 1992). Estimated emissions from landfilled waste are therefore subtracted from those produced by incineration (expressing methane emissions in terms of carbon dioxide equivalents, on the basis of global warming potentials) for comparison with emissions from average UK generating plant mix (table 4).

Emission	Emission savings (grammes of oxide per kWh)	Annual emission savings per typical 25 MW scheme (tonnes of oxide)
Carbon dioxide (equivalent)	1,800 ± 800	350,000 ± 150,000
Sulphur dioxide	7.3	1,400
Nitrogen oxides	1.0	190

Table 4: Emission savings relative to average UK plant mix in 1990 assuming current trends continue. Note: The figures for displaced emissions shown here assume that the alternative disposal route would be to a landfill site with gas collection and flaring, and that fugitive emissions of methane would therefore be avoided by incineration.

Concerns about emissions from incineration, particularly about heavy metals and dioxins, have produced strict emission regulations now embodied in European Union and UK legislation on incineration. This has led to the evolution of incinerator and gas cleaning systems that can meet these and even more stringent standards. Nonetheless incineration projects often still face opposition at the planning stage because of fears about the environmental effects of these emissions.

Waste incineration leads to:

- emissions of combustion products, which could include some acidic gases, heavy metals and some chlorine containing species, including very low levels of dioxins;

- incinerator ash, including the ash collected in precipitators etc. which can be enriched with heavy metals (particularly lead and cadmium);

- effluent disposal from any wet scrubbing processes included in the gas clean up system employed, boiler blowdown and cooling water circuits.

Digesting waste through anaerobic processing leads to:

- possible odour problems during storage and processing;

- the production of a compost byproduct which could contain some heavy metals or other contraries, the levels of which will vary from place to place, dependent on waste composition. The use to which these materials can be put (and thus their value) will be governed by the level of these contaminants, which may be restricted by regulation.

Locally, incineration reduces the requirements for landfill and may act as a spur to recycling when part of an integrated system.

E. Economics

The resource potential of these wastes and economics associated with their use will be significantly influenced by developments in environmental legislation and trends in disposal practice. These trends are difficult to predict with confidence, so two different views of the future have been considered. The first assumes that current trends continue and legislation currently proposed takes full effect; the alternative assumes a more environmentally sensitive future in which environmental constraints on waste disposal are more stringent and there is an increasing move to waste minimisation and recycling.

If current trends continue it is assumed that the following developments occur by 2025:

- improved standards for landfill push up costs of landfill disposal for MSW in real terms by £10/tonne from a range of £5 – £25 to £15 – £35/tonne;

- environmental, economic and public pressures favouring waste minimisation and recycling lead to an overall net reduction in household waste generation of 10%, with a greater reduction (20%) in the arisings of commercial and industrial waste;

- changes in the structure of the UK waste disposal industry, including the establishment of Local Authority Waste Disposal Companies (LAWDCs) will decrease the distinction between wastes generated in commerce and households, with the amount of commercial waste entering the MSW stream doubling (from one third of all arisings at present, to two thirds);

- residual commercial and industrial wastes will continue to be disposed of to landfill under short-term disposal contracts. However, their relatively high calorific value makes them an attractive feedstock, generally for small-scale on-site incineration. This contribution has not been modelled.

If more environmentally sensitive policies and practices are adopted it is assumed that:

- there are greater pressures against landfilling, which lead to an increase in MSW disposal costs of around £20/tonne to £25 – £45/tonne;

- there is a major national effort backed by public pressure towards waste minimisation and recycling, implemented in part by adoption of "sort at source schemes" for domestic waste, this results in a 20% reduction in household waste and a 30% reduction in commercial waste.

- changes in the structure of the waste disposal industry occur in the same way as if current trends continue;

- the costs of incineration and waste combustion are increased by 20% to allow for more stringent pollution control measures and the costs of disposal of incinerator ash are increased three-fold.

The trends in disposal practice under these assumptions are summarised in table 5. The corresponding MSW supply curves are given in figure 5. In addition, 50% of sewage sludge arisings (0.75 M tonnes/year) are assumed to be Accessible at between £0 and £20 per tonne dry solids.

	Year	House-hold Waste	Commercial / Industrial		MSW			
		Mtonnes/ year	Mtonnes/ year	% to MSW	Mtonnes/ year	Mtce*/ year	% to Incineration	% Resource Recovery
Current	1992	25	25	33	33.3	13.4	10	0
Current trends continue	2005	24	23	45	34.4	14.3	13.5	6.5
	2025	22.5	20	67	35.9	15.7	25	25
More environmentally sensitive policies and practices adopted	2005	23	22	45	32.9	13.7	20	13.5
	2025	20	17.5	67	31.7	13.8	25	35

Table 5: Summary of waste trends under different views of the future. Note: MSW assumed to consist of household waste plus stated percentage commercial/industrial waste. Calorific values (GJ/te) used: coal 27, household 9, commercial and industrial 16.5. *Mtce - million tonnes of coal equivalent

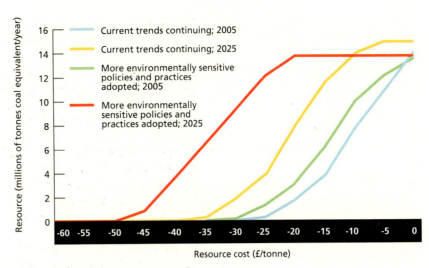

Figure 5: Municipal solid wastes resource

Capital and Operating Costs

Four different technical options for using these wastes as fuel are considered. Tables 6 to 9 give best available estimates of the capital and operating costs associated with these options for both views of the future.

Cost and performance data	Current trends continue	More environmentally sensitive policies and practices adopted
Power (export) (MW$_e$)	27	21.6
Capital cost (£ M)	79	95
Fixed operating costs: (£ M/year)		
Maintenance	1.82	2.18
Rates/insurance @ 1.5% capital	1.18	1.43
Other	0.67	0.67
Labour	1.5	1.5
Variable operating costs: (£ M/year)		
Residue disposal	1.76	5.28
Reagent and other costs	0.97	3.25
Project lifetime (years)	20	20
Construction time (years)	3	3
Nominal GCV (GJ/tonne)	10.8	10.8
Waste throughput (tonnes/hour)	56	56
Availability factor (%)	85	85
Conversion efficiencies (fuel to electricity) (kWh/tonne)	485	388
Additional system load (%)	not applicable	20

Table 6: MSW incinerator plant costs and performance data. Costs are for all new equipment and associated civils.

Cost and performance data	
Power (export) (MW$_e$)	1.0
Capital cost (£M)	9

Table 7 continued overleaf

Operating costs: (fixed and variable) (£M/year)	0.6
Project lifetime (years)	15
Construction time (years)	1.5
Nominal GCV (GJ/tonne)	6
Waste throughput (tonnes/MSW/hour)	11
Availability factor (%)	95
Conversion efficiencies (fuel to electricity) (kWh/tonne/MSW)	120

Table 7: MSW digestion plant costs and performance data. Costs are for all new equipment and associated civils.

Cost and performance data	
Power (export) (MW$_e$)	0.1
Capital cost (£ M)	0.35
Fixed operating costs: (£/year)	
Maintenance	2,000
Other	2,000
Labour	10,000
Variable operating costs: (£/year)	
Residue disposal @ -£29.50/tonnes fuel*	58,150
Project lifetime (years)	20
Construction time (years)	2
Nominal GCV (GJ/tonne)	18.9
Waste throughput (tonnes/hour)	0.25 dry solids
Availability factor (%)	90
Conversion efficiencies (fuel to electricity) (kWh/tonne)	560 dry solids

Table 8: Sewage sludge digestion plant costs and performance data. Costs are for all new equipment and associated civils. *Savings on effluent disposal. No credit assumed for compost.

Cost and performance data	Current trends continue	More environmentally sensitive policies and practices adopted
Power (export) (MW$_e$)	13.5	12.2
Capital cost (£ M)	39	42.9
Fixed operating costs: (£ M/year)		
Maintenance	0.79	0.87
Rates/insurance @ 1.5% capital	0.54	0.60
Other	0.69	0.69
Labour	0.97	0.97
Variable operating costs: (£ M/year)		
Residue disposal	1.28	3.84
Reagent and other costs	0.44	0.66
Other income	(0.62)	(0.62)

Project lifetime (years)	20	20
Construction time (years)	2	2
Nominal GCV (GJ/tonne)	10.8	10.8
Waste throughput (tonnes/hour)	27	27
Availability factor (%)	85	85
Conversion efficiencies (fuel to electricity) (kWh/tonne)	500	450
Additional system load (%)	not applicable	10

Table 9: Resource recovery plant costs and performance data. Costs are for all new equipment and associated civils.

In all cases the wastes could be used to produce electricity, heat or steam, or to operate a combined heat and power plant. The choice between these energy products will depend on the relative economics at an individual site, and particularly on the availability of local heat markets. While such markets may be available in a few cases, the principal use for the energy will continue to be for electricity production. The discussion here concentrates solely on electricity production.

Municipal Waste Incineration
Incineration systems are well developed and fully commercial in Europe, although it is many years since a system was built in the UK. Information on the costs of such systems is based on commissioned feasibility studies currently in progress for UK sites.

The costs depend very much on the scale of operation. The costs presented in table 6 are for a plant size of 56 tonnes/hour (400,000 tonnes/year). This is about the smallest size of plant which would be considered in the UK, and then only in regions where disposal charges are highest. Larger plants could be considered, but there are relatively few regions where the waste arisings would support such a development.

Biological Processing
Two cases are considered, the anaerobic digestion of MSW and of sewage sludge.

Table 7 relates to a waste processing plant which handles 11 tonnes of raw MSW per hour. Such a plant could produce 1.0 to 1.3 MW of electricity for export, along with a compost fraction suitable for lowgrade uses such as intermediate cover on landfills. The data is derived from an assessment of overseas experience.

Table 8 provides data for a typical sewage treatment plant handling 50,000 cubic metres/year of thickened sludge, equivalent to 0.25 tonnes/hour of dry solids. This is based on an analysis of existing UK plant.

Refuse-Derived Fuels
There is a wide spectrum of plants that could be designed to process waste, allowing recovery of some materials and rejection of inert or other undesirable components, with energy recovery from the remainder. The data in table 9 refers to a c-RDF plant with a throughput of 200,000 tonnes/year in which waste is screened and pulverised and used without drying or densification to produce energy in a combustor. This technical option requires demonstration in the UK but has been fully commercialised overseas.

F. Resource-Cost Curves

Resource-cost curves for these wastes are calculated by combining the information about the costs at which the feedstock would be available with the information on the capital and operating costs of the various options. Resource-cost curves have been compiled for the years 2005 and 2025 at 8% and 15% discount rates for a future assuming present trends continue and one assuming that more environmentally sensitive polices and practices are adopted.

A complication is that the various technologies will be competing for the same resource and will achieve conversion to electricity at different efficiencies. Estimates are therefore required of the amount of each waste treated by each technology. This must include the proportion going to landfill in order to provide resource data for the landfill module. These proportions are indicated in table 5. It is further assumed that large-scale incineration would only be considered for municipal waste arising from large urban conurbations; this would incur the highest disposal fees if disposal were via transfer station to long-distance landfill. This waste is assumed to be the highest value waste, a third of the arisings indicated in figure 5. The remaining two thirds of the waste will generally derive from more diffuse sources, such as smaller cities, towns and shire counties. These arisings may be amenable to utilisation through smaller-scale resource recovery schemes, either as c-RDF or d-RDF. Anaerobic digestion of MSW is included within resource recovery.

The energy systems modelling examines the time-varying contribution between 1995 and 2025 and it is therefore necessary to construct a time-varying version of figure 5 for each view of the future rather than the simple snap-shot resource-cost curves presented for 2005 and 2025 in that figure. For large-scale incineration this complication has been accommodated by considering only how the minimum price of the resource varies with time. This was taken as the price at which the most expensive one-third of the waste becomes available, in line with the assumption above. For RDF the characteristic price was taken as the mid-price of the remaining resource. These are conservative assumptions. The resource size and cost were then assumed to change linearly with time.

The assumption is made that all sewage sludge is assumed to be treated by anaerobic digestion. This could yield approximately 0.4 TWh/year of electricity. This contribution is not included in the analysis below.

The graphs presented in figures 6 to 9 are for the Maximum Practicable Resource which takes account of changes in the available resource with both time and view of future waste management policies and practices. These curves are consistent with the input data used in the energy systems modelling, the results of which are presented in Section H. The data assuming present trends continue was used for the modelling of the Composite, Low Oil Price, High Oil Price and Shifting Sands scenarios and that assuming more environmentally sensitive policies and practices were used for the modelling of the Heightened Environmental Concern scenarios.

The variation in the cost of electricity is due to the variation in the assumed gate fees for waste disposal. However the gate fee for any particular plant will be subject to competitive pressures and in cases where the calculation results in electricity costs below its market value it would be more realistic to do the calculation the other way round, taking the electricity price as given and calculating the gate fee required to cover costs. The discounted cash flow analyses over a 20 year plant life have been carried out at 8 and 15% discount rates. These rates may be considered low for capital intensive schemes such as energy-from-waste plant. However, developers are generally in competition with alternative low-cost disposal, mainly landfill, and therefore need to accept a lower return against a long-term waste disposal contract.

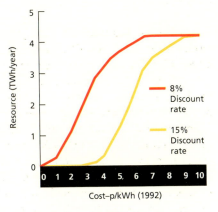

Figure 6: Maximum Practicable Resource-cost curves for municipal solid waste in 2005 assuming current trends continue.

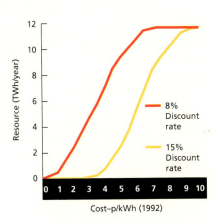

Figure 7: Maximum Practicable Resource-cost curves for municipal solid waste in 2025 assuming current trends continue.

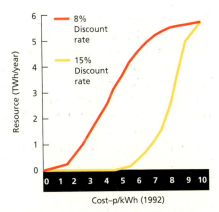

Figure 8: Maximum Practicable Resource-cost curves for municipal solid waste in 2005 assuming more environmentally sensitive policies and practices are adopted.

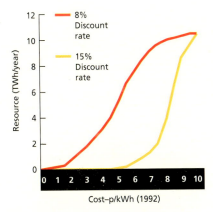

Figure 9: Maximum Practicable Resource-cost curves for municipal solid waste in 2025 assuming more environmentally sensitive policies and practices are adopted.

G. Constraints and Opportunities

Technological/System

The major techno-economic barriers faced by the various options differ considerably. Modern incineration technology is well developed in Europe and elsewhere, but there have been only limited incineration projects in the UK for over 15 years. Modern examples of plant in operation which are technically and financially successful and which fully meet today's more stringent emission regulations are badly needed to establish long-term confidence in the technology amongst the waste disposal community, financiers etc. The other technologies discussed above are generally less mature. Although most of the components that would be used in the systems are developed and have been used in other applications, the complete systems need to be put together and evaluated to provide a basis of confidence for future development and implementation of the technology. In practice completing the development and demonstration stages in the evolution of these technologies may prove difficult. The plant involved has to be similar to full scale if meaningful lessons are to be learnt. The companies currently involved in these developments in the UK often do not have the necessary resources to finance extended development work at larger scale. This has led to a general reliance on overseas technology and systems. However, incompatibilities and lack of experience can lead to premature projects entering the commercial market without adequate contingency. When technical difficulties are encountered with such projects both inadequate financial provision and a lack of contractual flexibility may hinder resolution of the problems.

Institutional

The increasing environmental pressures on waste disposal are on the whole acting to favour energy recovery from these wastes. Landfill will continue to be a major waste disposal route for the UK, but incineration and resource recovery are increasingly being considered as long-term options alongside waste minimisation and recycling as part of an integrated approach to waste management. However, changes in waste management regulation and environmental legislation have caused a good deal of uncertainty, particularly over costs, and this has inhibited the development of a concerted industrial effort to make use of wastes in the most effective way. A period of legislative stability would allow industry to formulate business plans to exploit these technologies and to establish the necessary long-term waste disposal contracts.

The uncertainties which stem from legislative and regulatory changes are exacerbated by uncertainties that surround the future value of materials and energy derived from these wastes. This is particularly inhibiting those technologies which are not yet ready for immediate commercial exploitation. Few companies are prepared to invest in development work that will lead to technologies ready for commercial deployment at a time when neither the income from waste disposal nor from energy sales can be predicted with certainty. The present NFFO arrangements have provided an immediate boost, particularly to incineration technologies, but this has largely resulted in the adoption of overseas technologies rather than the development of UK capabilities.

A major barrier to the rapid implementation of these technologies is the nature and experience of the UK waste disposal industry. The industry has until recently been principally concerned with landfill disposal; it is not familiar with the technology for energy recovery. The situation is even more acute where incineration and RDF technologies are concerned, although some specialist contractors are now developing expertise. There is also an encouraging level of interest being shown in these technologies by companies who do have experience of both technical and commercial facets of energy production, for example, some of the heat service companies and the privatised electricity companies, as well as some entrepreneurs who recognise that this area will provide a significant business opportunity in the future.

Financing projects in this area poses some problems. First there have been a number of difficult and unsuccessful waste incineration projects which has given the area a poor reputation. Some projects, for example a £50 to £100 M incineration plant, are large and organisationally complex with the economics dependent on many parameters and organisations. On the other hand, many projects tend to be small. Though organisationally simpler and technically less risky, these may be too small to attract the interest of finance houses. In some cases these problems have been circumvented by the establishment of joint ventures but this pattern is not yet well established.

H. Potential & Prospects – Modelling Results

Scenario	Discount rate	Contribution (TWh/year)			
		2000	2005	2010	2025
HOP	8%	3.7	4.2	4.6	6.1
	15%	1.1	0.52	0	0
	Survey (10%)	1.1	4.2	4.6	6.1
CSS	8%	3.7	4.2	4.6	6.1
	15%	1.1	0.52	0	0
	Survey (10%)	1.1	4.2	4.6	6.1
LOP	8%	3.7	4.2	4.6	6.1
	15%	1.1	0.52	0	0
	Survey (10%)	1.1	0.52	0	0
HECa	8%	3.0	3.45	3.7	4.3
	15%	1.1	0.52	0	0
	Survey (10%)	1.1	0.52	3.7	4.1
HECb	8%	3.0	3.45	3.7	4.3
	15%	1.1	0.52	0	0
	Survey (10%)	1.1	0.52	0	0

Table 10: MSW and resource recovery – potential contribution under all scenarios.

Environmental burden			Incremental contribution (TWh/year)			
	Constraint	Base case scenario	2000	2005	2010	2025
Carbon dioxide	- 5%	CSS	0	0	0	-5.0
	- 5%	HOP	0	0	-0.49	-61
Carbon dioxide	-10%	CSS	0	-3.6	-4.6	-6.1
	-10%	HOP	0	0	-4.6	-6.1
Sulphur dioxide	-40%	CSS	0	0	0	0
	-40%	HOP	0	0	0	0
Nitrogen oxides	-40%	CSS	0	-3.6	-4.6	-6.1
	-40%	HOP	0	-3.6	-4.6	-6.1

Table 11: MSW and resource recovery – incremental contribution under environmental constraints (TWh/year above base case scenario contribution).

Incremental contribution (TWh/year)				
Discount rate	2000	2005	2010	2025
8%	0	0	0	0
15%	0	0	0	6.1
Survey	2.7	0	0	0

Table 12: MSW and resource recovery – incremental contribution under Shifting Sands assumptions (TWh/year above CSS contribution).

The modelling results are given in tables 10–12. These show three key features. First, there is a significant uptake of incineration technologies (3 to 6 TWh/year) at the 8% discount rate. Uptake is not affected greatly by the assumptions on waste management policies and practices, with the greater waste disposal fees available in future scenarios being off-set by the cost of meeting higher environmental standards, particularly under more environmentally sensitive scenarios.

Second, uptake is highly sensitive to discount rate. This reflects the high specific capital cost of incineration, particularly at small scales of operation. This sensitivity is evident in many of the incineration schemes currently being developed under the 1990 and 1991 NFFO tranches, where the cost of financing is proving to be a common obstacle. At the 15% discount rate the model results predict no uptake beyond the current NFFO tranches. Specific opportunities will exist where disposal fees are higher than the minimum assumed in this modelling exercise. However, it is evident that these schemes will carry a greater price-sensitivity risk.

Finally, RDF technologies are not taken up under any scenario. This is a realistic result given the modelling assumption that the highest value wastes would be utilised exclusively through large-scale incineration. The result would be influenced greatly if the higher value wastes were accessed. However, this would have resulted in double-counting in this modelling. The result is also sensitive to the waste disposal credit available and, in particular, whether a 'green premium' is available to assist materials recycling. c-RDF is assumed to have an energy yield of 500 kWh/tonne, so that an additional disposal fee or 'green premium' of £10 per tonne would provide additional income equivalent to 2p/kWh. This would bring c-RDF into line with incineration at the largest scale of operation.

In practice, local factors will determine the mix of large and small-scale incineration and the uptake of RDF. Moreover, the analysis here is for electricity production only. Specific opportunities may exist for CHP or district heating which would influence this picture greatly. Also, the uptake of c-RDF as an integral part of resource recovery would be affected by measures to support the development of recycling.

The potential contribution from sewage sludge digestion was also modelled. It was found that a contribution of 0.4 TWh/year occurred in each scenario for all years and all discount rates. There was no change to the contribution under the shifting sands or environmental constraints assumptions.

II. Programme Review

A. United Kingdom

The range of technologies for recovering energy from wastes are at differing stages of maturity. Incineration technology is fully developed in Europe, but there is little recent successful history of implementing this technology in the UK. Landfill gas technology is now becoming well established, although so far only a small proportion of the available resource is being utilised effectively. The alternative technologies RDF and biological processes are not yet well established.

The major barriers to the commercialisation and full deployment of these technologies in the UK can be summarised as follows.

- There are major uncertainties about future environmental legislation concerning waste treatment, with regulations both about landfill practice and incineration under review. These uncertainties discourage both investment in commercial projects to exploit wastes today, and are even more of a disincentive to industrial involvement in research and development work that could lead to commercial projects in five to ten years time.

- The UK waste disposal industry is also undergoing major structural changes, due in part to the changing roles of the public sector waste disposal authorities. In addition, the existing industry has been principally concerned with disposal to landfill and is not familiar with either the technical or commercial aspects of energy production or marketing. NFFO has stimulated some companies with this experience to enter the area, but strong alliances having a capability both in waste management and the energy technologies concerned have yet to develop.

- The technology involved in using these wastes is viewed as technically difficult and risky to those organisations without experience. A number of costly failures in the past have reinforced this view. This means that it is difficult for companies to finance field trial and demonstration projects in a way that allows sufficient capital to be available to tackle any problems that arise. The industry is also not technically or financially strong enough to finance pilot or development work. This can lead to premature full-scale commercial schemes which are prone to fail.

The Government's Programme

The use of municipal wastes as fuel is principally dealt with within the Department of Trade and Industry's Biofuels Programme, but close links have been established between this work and the Department of Environment's programme on waste disposal.

The main aims of DTI's work in this area are as follows:

- to encourage the adoption of technologies to use wastes as fuel where these are economic and environmentally acceptable;

- to develop improved technologies where this is a prerequisite to commercial exploitation;

- to assess the wide range of available technical options taking full account of all relevant energy, environmental and economic criteria.

The main activities in the current programme include the following:

Waste combustion
- evaluation of the costs and benefits of incineration;

- evaluation and identification of promising novel approaches to incineration being tried in the UK or overseas; with feasibility studies for incineration schemes.

- evaluation of the costs and benefits of a range of RDF processes.

Biological Processes
- evaluation of a range of concepts under development world-wide, to identify those of relevance to the UK;

- support for the pilot-scale development of systems involving digestion of waste with a high solids content.

This work has benefited from good links with work going on internationally established through the IEA Bioenergy Agreement. Particular benefit has been derived from the Bilateral Agreement with the USA, where close links have been established both with work in the Government Programmes and with industry.

B. Other Countries

As in the UK, the utilisation of municipal wastes overseas depends on the cost competitiveness with other energy sources and alternative treatment and disposal options, usually landfill. Considerable differences in national approach are evident and have led to a range of uptake and interest in technological development. The UK, with its ample supply of suitable low-cost landfill capacity and comprehensive national energy supply, is generally regarded as well down the league in all but landfill gas exploitation. In particular, differences in energy pricing, taxation and/or use of district heating have made municipal and industrial wastes an attractive feedstock in certain countries, particularly in Scandinavia and Northern Europe (70% utilisation in Denmark, 55% in Sweden). At the same time, concern over uncontrolled emissions from landfills has prompted several European countries to impose limits on the organic content of wastes destined for landfill, necessitating a new round of investment in incineration and other pre-treatment options. For example, The Netherlands, faced with an acute shortage of suitable landfill capacity, has adopted ambitious targets to halve landfill requirements for priority wastes by the year 2000 by a combination of methods, including a ban on organics and a 250% increase in incineration. Japan, faced with even more acute landfill limitations, is pressing to increase incinerator capacity to deal with 80% of MSW arisings – and is even adopting ash melting techniques to further reduce landfill requirements.

The above, however, paints a distorted and over-simplified picture. What is evident from studying national policies is that the debate has moved on from optimal disposal to the integration of waste treatment with waste minimisation and recycling as part of more environmentally responsive waste management. Within this there is considerable on-going debate particularly as to the role of incineration. The principal response has been to progressively tighten environmental standards, thus increasing treatment and disposal costs. Other pre-treatment options are increasingly being considered alongside incineration as their environmental pressures increase and the costs of landfill rise. The options certainly include source separation and centralised recycling (e.g. MRFs), separation and composting/anaerobic digestion of organic wastes and incineration. At present some 20 municipal waste-based anaerobic digestion schemes and 30 incineration/resource recovery schemes are in final planning or construction in Europe. The manufacturers involved will gain considerable market advantage from these developments. Unfortunately, UK manufacturers have had little involvement.

The DTI programme has been informed of overseas developments by an on-going series of reviews and studies focused mainly on individual systems and technologies. The DTI has also subscribed to the International Energy Agency activity in MSW. This provides information exchange between twelve participating countries on various aspects of MSW utilisation, including landfill gas, incineration, waste processing, sampling and analysis, anaerobic digestion and ash management. This is now providing a major complementary source of information and contact with national programmes.

Figure 5: The Saint Ouen Waste-to-Energy Plant, Paris. (Courtesy TIRU.S.A. for SYCTOM).

Landfill Gas

I. Technology Review

A. Status of the Technology

Under the anaerobic (oxygen free) conditions of landfill sites, organic waste is broken down by micro-organisms, leading to the formation of landfill gas. Landfill gas is primarily a mixture of carbon dioxide and methane (in roughly equal quantities), with a large number of trace components such as halogenated hydrocarbons and organosulphur compounds. The methane content of the gas (typically around 40 to 60% by volume) accounts for its potential use as a fuel.

Figure 1: BFI's Packington landfill site unloading refuse.

Figure 2: Tarmac Econowaste's Rowley Regis quarry.

Landfill gas is collected through a series of gas wells on which a small suction pressure is applied. A wide variety of designs of gas wells and collection systems are available. The choice will depend to some extent on site-specific factors, such as depth of waste and water table.

Gas collection for energy recovery can often complement environmental protection measures now in force. There is a potential risk to the local and the global environment from the escape of landfill gas, and its control is often required under strengthened environmental protection measures at modern landfill sites. Energy recovery from landfill gas can follow once the necessary control has been achieved. A well designed landfill site and gas collection system can ensure integration of these two activities.

Figure 3: Yorkshire Brick's restored landfill site.

Energy recovery from landfill gas began in the UK in the late 1970s. The early days saw landfill gas collected from sites which were adjacent to consumers of heat (brick kilns were prime examples) and used to displace a conventional fuel. This approach to using landfill gas, however, is limited by the availability of customers sufficiently close to a landfill site to make the sale of the gas economic. There still remain some suitable heat loads which could be supplied from nearby landfills – and these should not be discounted when assessing the potential for energy recovery – but more and more projects are now using landfill gas as a fuel to generate electricity.

The use of landfill gas as a fuel for electricity generating plant began in the UK in the mid 1980s. The common types of engines used are gas turbines, dual fuel (compression ignition) engines and spark ignition engines. Engine sizes available range from a few hundred kW to several MW. Typically the smaller engines tend to be spark ignition and the larger, gas turbines. Fuel conversion efficiency for the generating sets can range from 26% (typically for gas turbines) to 42% (for dual fuel engines).

The choice of engine depends on many factors, including the gas generation rate and its composition, the plant efficiency, plant availability, necessary gas pre-treatment, maintenance requirements, operators' familiarity with the plant, plant flexibility and life expectancy and, not least, cost. Of the 37 electricity generation projects existing in the UK at the end of 1992, 28 are based on spark ignition engines, seven on dual fuel engines, and two on gas turbines. Life expectancy depends on how the plant is operated, but is likely to be between 10 years for the higher speed spark ignition engines and 20 years for the slower speed dual fuel engines. The gas yield will depend on the nature of the landfill. For a large modern

Figure 4: An internal combustion engine fuelled by landfill gas at Springfield Environmental's Appley Bridge site.

landfill, usable landfill gas may be generated for between 15 and 30 years.

Landfill gas can be used as a boiler fuel in conjunction with steam turbine generating plant but the large scale usually required for economic steam plant limits the potential for landfill gas as a fuel. Moreover steam plant has a relatively poor thermal efficiency compared with competing technologies.

For most applications, landfill gas is used without extensive purification, other than removal of moisture and particulate matter. For some electricity generation projects further cleaning is undertaken to limit the risk of corrosion, erosion, wear, deposit build-up or trace gas emissions from plant. More extensive clean-up is needed if landfill gas is to be used as a substitute natural gas (e.g. injecting it into conventional gas pipelines, or as compressed gas in applications such as a vehicle fuel, or through liquefaction to liquid natural gas). Other applications, such as using landfill gas as a feed stock for fuel cells for electricity generation or for chemical production (e.g. methanol), require an even higher standard of purification.

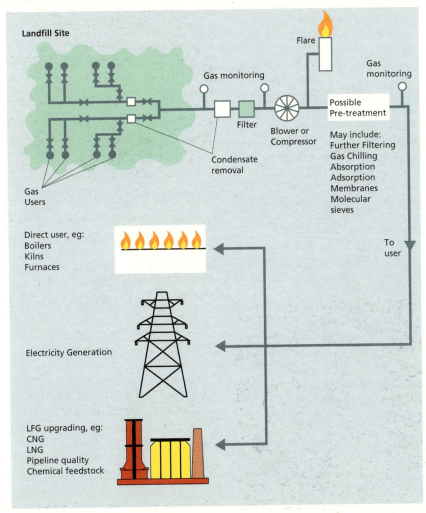

Figure 5: Typical landfill gas abstraction layout.

B. Market Status

Landfill gas is often cited as one of the more developed renewable energy sources. It is used as a commercial fuel in many countries around the world and it is supporting the expansion of an industry dedicated to its collection and use.

The potential of landfill gas as an energy resource is inextricably linked with waste management practices. As standards for waste disposal have increased (leading to better engineered sites), so the gas generation potential has risen. Deeper sites, fully contained, provide a better environment for the activity of the methane-producing bacteria than shallow, poorly-capped sites. As landfill technology becomes better understood the environmental risks associated with it have become more apparent. This has led to a strengthening of the legislation governing landfill gas control. Energy recovery can be an integral part of this control and offset some of the costs associated with it.

Figure 6: Areas where landfill gas is exploited in the UK as at January 1993.

Exploitation of landfill gas as a fuel has been encouraged through the introduction of NFFO. The guaranteed market for electricity provided by NFFO, at a premium price, has enabled a large number of projects to be brought on stream which otherwise would have been at best only marginal and therefore difficult to finance. However, uncertainties over the future development of landfill gas still remain and a number of technical, commercial and institutional issues stand in the way of its full commercial exploitation.

Project developers include landfill operators, independent energy firms (including some Regional Electricity Companies) and a few industrial entrepreneurs.

Current status	Commercial – with limitations
UK installed capacity	Approximately 75 MW
World installed capacity	Approximately 500 MW
UK industry	Several well established UK equipment suppliers

Table 1: Status of technology (electricity generation).

C. Resource

The potential for energy recovery from landfill gas depends on waste management policy and practices. If waste management policy drives organic waste to alternative waste management techniques such as incineration or composting, then this will affect the amount of landfill gas available for use as a future fuel. Currently in the UK about 90% of municipal solid waste is deposited in landfill. In the foreseeable future landfill will remain the major waste disposal option. The trends towards a better understanding of landfill sites and improvements in their design and operation are expected to increase the potential for energy recovery.

It is generally accepted that energy recovery from landfill gas is only possible for those sites sufficiently large to sustain substantial gas generation. This minimum size is often taken to be around 200,000 tonnes of waste in place, deposited over the previous 10 to 15 years. The Accessible Resource is therefore limited to the number of sites that fall within this category. The current potential for energy recovery from LFG is about 2.5 million tonnes of coal equivalent per year in England and Wales, which equates to approximately 5 TWh/year of electricity. This assessment was based on a survey of landfill sites categorised according to waste composition and design using an empirical landfill gas prediction model to provide the National Assessment for energy recovery (ETSU Contractor Report B 1192 provides full details). The potentials for Scotland and Northern Ireland have been estimated using waste arisings data, assuming the gas generation potential per tonne of waste to be the same as England and Wales.

Discount rate	England and Wales	Scotland	Northern Ireland	UK
8%	5	0.3	Nil	5.3
15%	5	0.3	Nil	5.3
25%	5	0.3	Nil	5.3

Table 2: Current Accessible Resource in the UK at less than 10p/kWh (TWh/y).

D. Environmental Aspects

The land needed for landfill gas power generation schemes is mainly that required for buildings to house the engine/generator sets, pumps, compressors and gas cleaning equipment. Engine/generator sets up to about 1 MW$_e$ are commonly supplied in portable containers which rest on concrete slabs. Typically these have a ground area about 25 sq.m and a height of about 3 m. Larger installations are usually housed in purpose-built buildings containing several engine/generator sets. Overall plant installations (including gas extraction plant and electricity connection equipment) may be around 25 m x 25 m in size. Landscaping and design minimise visual intrusion, which is not expected to be a significant factor because of the relatively small structures involved. Some noise is generated from the gas compressors, engine and exhaust system. With adequate insulation, siting and design, noise levels can be kept to acceptable limits. The characteristics of reciprocating engine (dual fuel and spark ignition) noise makes attenuation more difficult than in the case of gas turbines, but even so, there are numerous examples of low noise limits being met by all types of engine. Water use is low, being limited to site services and occasional cooling water make up. Ecological impacts of landfill gas exploitation are not expected to be significant. Transportation impacts will be minimal after construction and will not add significantly to vehicle movements on an operational landfill.

To meet more stringent regulation and control of landfill gas, site operators can choose from a range of options:

- install migration barriers and gas vents;

- collect landfill gas and flare it;

- collect landfill gas and recover energy from it.

Under the first two options no energy benefit is obtained. Under the first there is the risk of odours and the emission of potentially hazardous gases in trace quantities. Moreover, there is a significant loading of methane in the atmosphere; methane is estimated to have a global warming potential per unit release many times that of carbon dioxide. Under the second and third options odour problems are minimised and the methane content of landfill gas is converted to the much less harmful carbon dioxide. There will be slight differences in other gaseous emissions (e.g. nitrogen oxides) due to the different combustion characteristics of flares and engines, but overall the net gaseous emissions from engines are unlikely to have more detrimental effects than those from flaring. Estimated stack emissions and environmental burdens from a LFG fired power generation set are listed in table 3.

Environmental burden	Specification (units per kWh$_e$)
Gaseous emissions:	
Carbon dioxide	1.3 kg[a]
Sulphur dioxide	0.3 g[b]
Nitrogen oxides	2.1 g[c]
Carbon monoxide	2.7 g[c]
Volatile organic compounds	2.1 g[c]
Particulates	0.07 g[b]
Heavy metals	Trace
Ionising radiation	None
Solid and liquid waste	Insignificant
Thermal emissions	2.3 kWh[d]
Amenity & Comfort:	
Noise	Minimal
Visual intrusion	Unlikely to be significant
Use of wilderness area	None
Other	No additional impact

Table 3: Environmental burdens associated with landfill gas exploitation.

Notes   a: About 50% of this carbon dioxide comes from the landfill directly as carbon dioxide
        b: Measured
        c: Emission at limit for German legislation
        d: Thermal emission is unlikely to be significant due to small scale (some could be recovered as useful heat).

The emissions given in table 3 are likely to be bettered in future by further development of the LFG engines.

Table 4 indicates that electricity generated from LFG displaces all of the emissions associated with average UK plant mix (1990). For the purposes of this analysis it is assumed that LFG not used to generate electricity would be flared off, with no net effect on emissions. Fugitive emissions of methane from the landfill would occur whether or not energy is recovered and therefore are not included.

Emission	Emission savings (grammes of oxide per kWh electrical)	Annual emission savings per typical 3 MW$_e$ scheme generating 22 GWh/year (tonnes of oxide)
Carbon dioxide	734	16,200
Sulphur dioxide	10	220
Nitrogen oxides	3.4	75

Table 4: Emissions savings relative to average UK plant mix in 1990.

E. Economics

Spark ignition engines, 600 kW to 3 MW		
Cost data		
Capital:	£790/kW	
Prime mover and generator	£510/kW	
Civils	£90/kW	
Gas collection	£80/kW	
C&I, switchgear and power connection	£110/kW	
Commissioning and installation costs, decommissioning and other costs are included in the above estimates		
Annual recurring costs:	£130/kW	
These are fixed costs, and include O & M for gas collection (£20/kW) and energy recovery plant (£110/kW)		
Performance data	Plant availability	95%
	Average load factor	88%
	Type of operation	base-load
Other	Construction time	1 year
	Lifetime	10 years

Table 5: Cost and performance data. Load factor is independent of season and has no diurnal variations.

The example plant chosen to assess the cost of electricity generation from landfill gas is a 2 MW (electric) power station, consisting of four 500 kW spark ignition engines. The capital costs include a proportion of the costs associated with the installation of a gas collection system, basic landfill gas clean-up (limited to water removal and filtering), the engine-generator sets, and associated electrical connection equipment. Annual costs are fixed and include O&M on both the gas collection and the electricity generation plant. This type of system is currently favoured by project developers; the majority of existing projects are based on this type of technology.

The costs have been based on the EEO New Practice Programme project 'Electricity Generation using Landfill Gas' (New Practice final report 19), a demonstration project at Shanks and McEwan's Stewartby landfill site. The costs in that report have been scaled up to represent a 2 MW plant, at mid-1992 prices, and modified slightly to incorporate cost data from sources other than relevant published case studies.

F. Resource-Cost curves   As the future potential for energy recovery depends on waste management practices and policy, two possible futures have been considered. The first assumes that current trends continue, but that substantial changes in the waste management industry still occur. The assumptions are:

- waste minimisation will still have an effect, but to a lesser degree than under the former scenario;

- by 2025, 50% of municipal solid waste (MSW) and 60% of general industrial waste (GIW) will go to landfill (cf. 90% and 100% today);

- landfill gas collection will be mandatory at 80% of landfill sites;

- there will be some improvements made in terms of gas collection due to developments in technology.

The alternative future is more environmentally sensitive; it assumes that substantial changes occur in the waste management industry as more environmentally sensitive policies and practices are adopted. Assumptions include:

- waste minimisation and recycling will have a significant downward pressure on waste requiring disposal (countered to some extent by increases in GDP);

- by 2025, it is assumed that only 40% of MSW and 60% of GIW will go to landfill;

- landfill gas collection will be mandatory in all landfill sites in the year 2000;

- there will be improvements in both the yield of gas from deposited waste and the efficiency with which it is collected, thanks to an improved understanding of the phenomena involved and development of the technologies used.

These two views result in different estimates of the Accessible Resource. Under both, the rate at which the recovered resource approaches the Accessible Resource will depend on the market for electricity (including future NFFO-type support) and the degree of interest shown by the industry. Further assumptions have been made about both of these issues in the derivation of the graphs for the Maximum Practicable Resource presented below.

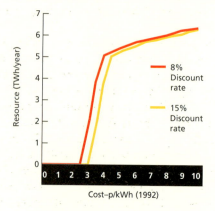

Figure 7: Maximum Practicable Resource-cost curves for landfill gas in 2005 assuming current trends continue.

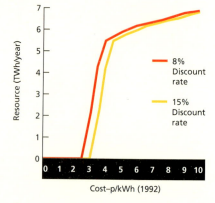

Figure 8: Maximum Practicable Resource-cost curves for landfill gas in 2025 assuming current trends continue.

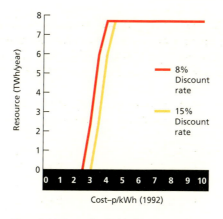
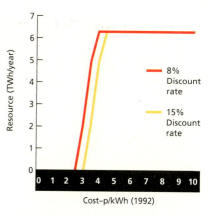

Figure 9: Maximum Practicable Resource-cost curves for landfill gas in 2005 assuming more environmentally sensitive policies and practices are adopted.

Figure 10: Maximum Practicable Resource-cost curves for landfill gas in 2025 assuming more environmentally sensitive policies and practices are adopted.

The data assuming current trends continue was used for the modelling of the Composite, Low Oil Price, High Oil Price and Shifting Sands scenarios and, assuming more environmentally sensitive policies and practices are adopted, was used for the modelling of the Heightened Environmental Concern scenarios. The results of the modelling exercise are presented in Section H.

G. Constraints and Opportunities

Technological / System

Exploitation of landfill gas for electricity production is based on established and proven technology, adapted to meet the fuel characteristics of landfill gas. Landfill gas has a relatively low calorific value (about half the heat value of natural gas) and it contains trace impurities which cause corrosion and accelerate wear. There are currently over 15 companies in the UK manufacturing or supplying landfill gas engines and many more companies who provide landfill gas collection and control systems. The most popular plant in the UK is based on reciprocating engines, either spark ignition (SI) or dual fuel compression-ignition engines using diesel oil as a pilot fuel. These engines are typically multi-cylinder units originally developed for operation on other gaseous fuels as stationary engines or for ship or heavy vehicle propulsion. Gas turbines are also used in some applications. The use of wear-resistant materials, development of special lubricants and careful attention to maintenance has increased the reliability of landfill gas-fuelled engines, and load factors in excess of 85% (including maintenance outages) are common. Connections to the local Regional Electricity Company (REC) grid are usually free of significant technical difficulties, although for smaller landfill gas power generation projects (< 1 MW) the cost of grid connection can be a significant burden, making a potentially viable project uneconomic.

Gas production usually begins within two years of waste emplacement as anaerobic conditions are established. Production reaches a maximum within a further five years, and then begins to decline after about 15 years. Although the gas may continue to be produced for several decades, gas quality and quantity are unlikely to be sufficient to sustain an electricity generation scheme for more than about 15 years (which may coincide with the life of the generating plant). The gas yield of a site is affected by many factors (such as waste composition, quantity, site capping, water content) and short term fluctuations caused by changes in barometric pressure occur over a time scale of hours and days. Uncertainty in the prediction of gas yields is therefore a major deterrent to potential developers and their financiers. Further developments in modelling and site assessment should help to reduce this uncertainty. In addition, there is scope for improvement in gas collection technology.

Institutional

NFFO has stimulated exploitation of landfill gas for power generation. There are now over 40 schemes having a combined DNC of over 70 MW_e supported under NFFO 1 and 2. Considerable experience has been built up and there is now a small but active industry comprising manufacturers, suppliers and consultants. Exploitation of landfill gas is generally complementary to environmental protection measures. The strengthening of regulations governing the control of landfill gas may lead to greater opportunities for exploitation. Possible changes in waste stream composition as a result of recycling and package reduction measures are likely to impact upon the future potential for landfill gas, as are possible limits governing the organic content of waste destined for landfill. Future European Union legislation covering emissions from stationary engines must take account of the energy and environmental benefits of landfill gas utilisation if these benefits are to be preserved. The development of landfill gas as an energy resource has been heavily influenced by the structure of the waste management industry in the UK. The provisions of the Environmental Protection Act (EPA) of 1990 have forced major changes in the industry in the UK. Before this law was passed, however, the industry was structured with both private companies and Local Government bodies (Waste Disposal Authorities – WDAs) operating landfill sites. Spending and borrowing restrictions faced by Local Government resulted in few being able to fund landfill gas exploitation schemes, so the majority of such projects in the UK have been initiated by private waste disposal companies together with a few entrepreneurial organisations established specifically to develop landfill gas schemes.

Prior to the implementation of the EPA the WDAs were responsible for regulating all sites, although they also operated sites along with the private sector. This situation is changing. WDAs are required under the EPA to relinquish operation of active landfill sites (i.e. those still accepting waste) either by selling them to the private sector, by setting up Local Authority Waste Disposal Companies (LAWDCs), or establishing joint ventures between the two. LAWDCs must compete for waste disposal contracts with other waste management companies, but are now free from the capital spending constraints faced by the former WDAs. LAWDCs can thus consider investing in landfill gas schemes. This restructuring of the industry therefore may provide new opportunities for establishing energy recovery projects at sites where such development has been difficult in the past.

Health

There are no significant hazards particularly associated with the use of landfill gas as a fuel. Safety and health hazards are those associated with operating conventional generating plant running on gaseous fuel.

H. Prospects – Modelling Results

Scenario	Discount rate	Contribution (TWh/year)			
		2000	2005	2010	2025
HOP	8%	0.54	0.27	0.12	0
	15%	0.54	0.27	0.12	0
	Survey (10%)	0.54	0.27	0.12	0
CSS	8%	0.54	0.27	0.12	0
	15%	0.54	0.27	0.12	0
	Survey (10%)	0.54	0.27	0.12	0

Table 6 continued overleaf

LOP	8%	0.54	0.27	0.12	0
	15%	0.54	0.27	0.12	0
	Survey (10%)	0.54	0.27	0.12	0
HECa	8%	6.7	7.7	8.6	6.3
	15%	6.7	7.7	8.6	6.3
	Survey (10%)	6.7	7.7	8.6	6.3
HECb	8%	6.7	7.7	8.6	6.3
	15%	6.7	7.7	8.6	6.3
	Survey (10%)	6.7	7.7	8.6	6.3

Table 6: Landfill gas – potential contribution under all scenarios.

Environmental burden			Incremental contribution (TWh/year)			
	Constraint	Base case scenario	2000	2005	2010	2025
Carbon dioxide	- 5%	CSS	0	0	0	6.9
	- 5%	HOP	0	0	0	6.9
Carbon dioxide	-10%	CSS	0	7.7	9.1	6.9
	-10%	HOP	0	0.70	8.4	6.9
Sulphur dioxide	-40%	CSS	0	0	0	0
	-40%	HOP	0	0	0	0
Nitrogen oxides	-40%	CSS	0	0	9.1	6.9
	-40%	HOP	0	0	0	6.1

Table 7: Landfill gas – incremental contribution under environmental constraints (TWh/year above base case scenario contribution; survey rate).

Incremental contribution (TWh/year)				
Discount rate	2000	2005	2010	2025
8%	6.3	6.1	-0.12	0
15%	4.6	4.5	-0.12	6.9
Survey	6.3	6.1	-0.12	6.9

Table 8: Landfill gas – incremental contribution under Shifting Sands assumptions (TWh/year above CSS contribution).

Table 6 presents the expected contributions from landfill gas under a number of scenarios. Under the CSS (Composite), LOP (Low Oil price) and HOP (High Oil Price) scenarios, fossil fuel derived power is at such a low cost that it prevents any additional uptake of landfill gas (the MARKAL model does not include any possible future NFFO orders). The cost of landfill gas-derived power may reduce somewhat through further requirements to collect landfill gas to mitigate methane emissions. Even so the capital cost and high operating cost of the power generating plant may still be too high for landfill gas to compete against, for example, coal-derived power or CCGT-derived power. For this reason, there is a decline from the current committed installed capacity to zero in 2025. The uptake is not sensitive to discount rate because the capital costs are assumed to have been already paid off for this installed plant. Under the Heightened Environmental Concern (HEC) scenarios all the Maximum Practicable Resource is taken up.

Table 7 suggests that under an additional 5% carbon dioxide reduction per decade constraint over the base case scenario there will be little effect on the uptake of landfill gas until the long term (2025). Under a higher constraint on carbon dioxide emissions (10% reduction per decade), however, landfill gas exploitation becomes more attractive due to its net carbon dioxide emissions benefits.

Landfill gas power generating projects have a relatively short lifetime, typically 10 years. The output therefore is sensitive to the large swings in energy price assumed under the Shifting Sands scenario. Other technologies with long plant lifetimes and therefore relatively small marginal costs will keep generating electricity at relatively low cost.

Factors which govern the adoption of landfill gas generating schemes are:

- capital cost, of both gas collection plant and energy recovery plant;

- operating costs, similarly separated out;

- plant lifetime;

- plant load factor;

- plant efficiency;

- industry inertia;

- waste management practices in terms of both waste management options and landfill practices.

The Maximum Practicable Resource-Cost data presented in figures 7 to 10 gives an indication of the anticipated range of costs that will be achieved depending on the waste management future and discount rate assumed. For example, it is expected that under the more environmentally sensitive waste management future, the cost of electricity from landfill gas will lie between 3 and 4.5 p/kWh for a discount rate of 15% in mid-1992 money.

II. Programme Review

A. United Kingdom

By January 1993, there were 50 existing landfill gas exploitation projects in the UK, with at least a further 15 planned or under construction. The use of landfill gas as a fuel for the generation of electricity or as a source of heat has proved to be technically feasible and commercially successful. Recent rapid development of the industry is a response to the opportunities provided by NFFO. For those projects supported by NFFO it has provided a guaranteed market for the sale of landfill gas-derived electricity. Through the premium prices paid, it has given a stimulus to the industry by making viable projects which would otherwise be only marginally economic. However, the existence of some electricity generation projects outside of NFFO support mechanism, in addition to the direct-use projects that have been developed without its support, indicates the proximity of landfill gas to true commercial viability as an energy resource.

Despite the successful development of the industry in the past 10 years, which to a large degree is a result of the Government R,D&D programme coupled with the enthusiasm and dedication of the industry itself, there remain significant barriers to the uptake of this resource. These include the following.

- Although some LFG schemes are currently commercially viable without market support, for most the market price for the energy they produce is still below that needed for commercial self-sufficiency.

- Availability of finance for LFG projects is affected by the risks perceived by financiers in the performance of equipment and predictions of resource size, the relatively small financing requirements of such projects (usually below £3M) and inexperience of prospective developers in obtaining finance and meeting the lender's criteria. This reduces the utilisation of the resource.

- Some sites face difficulties in accessing the electricity market through the high costs of connection to the REC grid. Access to suitable heat loads limits the uptake of LFG for direct use or CHP.

- Concern over the effects of possible future environmental protection legislation, such as pollution abatement regulations governing emissions from LFG combustion, acts as a further barrier to implementation.

The overcoming of these barriers is dependent on a co-ordinated approach by industry and Government. Within the growing landfill gas industry, organisations are involved in their own development programmes. These are highly market-orientated and tend to be focused on product development to maintain an organisation's competitive edge in the burgeoning market. Government can contribute to the move to commercialisation of landfill gas by providing a focus for the industry programmes and by addressing the generic technological barriers and the institutional obstacles.

The Government's Programme
The DTI's work on energy recovery from landfill gas forms part of the Energy-from-Waste Programme, which in turn is part of the overall Biofuels Programme. The aim of the landfill gas programme is to encourage the adoption of technologies that use landfill gas as a fuel, where it is economic and environmentally acceptable.

The main activities in the current programme include:

- assessment of the national landfill gas resource;

- commercial demonstration of a range of techniques for abstracting, treating and using landfill gas for electricity and heat production;

- promotion of the results of these projects and other examples of good landfill gas management and utilisation practice;

- development of improved techniques for controlling, managing and estimating yields of landfill gas;

- research at both pilot scale and in the laboratory to improve the understanding of the processes that lead to landfill gas production.

This work is carried out in close co-operation with the Department of Environment's (DoE) landfill technology R&D programme and close links are maintained with other government agencies around the world. Examples of such links are the IEA Bioenergy Agreement (Task XI – MSW Energy Conversion), under which Sweden, Norway, Denmark, the Netherlands, Canada, the USA and the UK collaborate and the UK/USA Bilateral Agreement on Biofuels research involving the DTI, DoE, US Environmental Protection Agency and the US Department of Energy.

B. Other Countries

Energy recovery from landfill gas is now world-wide. The benefits of using landfill gas have been recognised in most countries where waste disposal practices result in significant amounts of landfill gas being generated. For countries now developing waste disposal plans incorporating landfill techniques, energy recovery from landfill gas forms a major part of an integrated approach to waste management. Developments in these countries have been based to a large extent on the experiences and practices of those countries who have pioneered the technology. The world leaders in the field continue to be the USA, Germany and the UK.

Co-ordinated research programmes by both industry and Government have contributed to developments around the world. Laboratory scale research into micro-biological activity, large scale field gas enhancement work and the development of large engines designed to run specifically on landfill gas have all received Government support. Pressures for greater understanding of the phenomenon of gas generation and for the development of improved technology to put it to greater use have come from both the environmental control and the energy management arms of Governments.

There is a good flow of information around the world through formal agencies such as the IEA, through close dialogue between experts and open access to published literature. This ensures that research funds from whatever source continue to be put to the best practicable use.

Specialised Industrial Wastes

I. Technology Review

A. Status of the Technology

Industry produces a diverse range of wastes which require disposal. Many of these have some inherent energy value and, if burned as a fuel, the total resource would have an energy potential equivalent to some 3.5 million tonnes of coal per year.

Very little use is presently made of the resource. This is generally because wastes arise in small quantities from specific operations in discrete locations and are regarded by industry as a nuisance to be dealt with at the lowest cost with minimum inconvenience. In the UK this usually means direct disposal to landfill through a private waste contractor. Incineration, where used, is principally adopted to cut disposal costs rather than to produce useful energy.

Figure 1: Californian tyre dump: a potential source of fuel for combustion and energy recovery.

Currently deployed incinerator technology was developed largely in the 1960s from small-scale packaged 'destructors' designed for batchwise sanitary disposal of residues. Typical plants operate at a few hundreds of kilograms per day and consist of a simple fixed hearth within a refractory-lined combustion chamber, often with a secondary chamber or afterburner for smoke control. Heat recovery, where incorporated, is effected in a subsequent step, generally using a fire-tube waste heat boiler. There are numerous UK and overseas manufacturers of small-scale plant suitable for the burning of commercial wastes.

Refinements to the basic technology may include automatic or continuous feeding, automatic de-ashing, advanced grate technologies and higher pressure water-tube boilers geared to electricity production. Larger scale systems have also been developed overseas in response to local factors, principally the high cost of landfill disposal and stringent environmental legislation. Neither of these have been significant driving forces in the UK and currently no UK manufacturer offers such plant at scales much above one tonne per hour (3 to 5 MW thermal). However, an increasing range of technologies, including advanced rotary kiln incinerators and fluidised bed combustors, are marketed by a number of internationally active equipment suppliers. Plant sizes range up to 5 tonnes per hour, with multiple units used to increase capacity above this.

The technology is currently undergoing rapid evolution, largely as a result of modern requirements for comprehensive emissions control. It is expected that standards will continue to be tightened progressively, necessitating further development of the technology. Requirements introduced under the Environmental Protection Act 1990 are adding significantly to the cost and complexity of modern systems. The trend in utilisation is therefore expected to be towards larger scales of operation, where the economics can be advantageous especially for wastes which can attract a substantial disposal credit.

Inherent in this trend is a move away from incineration as a means of onsite disposal to offsite contractor operation, relying on waste from a number of sources to provide adequate feedstock. Such commercial developments tend to face strong local opposition and are likely to have strict environmental and planning controls applied to them. In addition to meeting statutory requirements, the physical design of buildings, traffic movements, noise and nuisance mitigation require careful consideration. Sympathetic siting is required to minimise the perceived impact of the development.

At larger scales of operation interest is also moving towards more advanced combustion technologies, including thermochemical conversion to provide a biogas or bio-oil feedstock and co-firing with other fuels on conventional large scale power-raising boilers. Demonstration and commercial schemes for a number of advanced technologies exist overseas, particularly for biomass and agricultural wastes. Such systems may offer significant advantages when applied to industrial wastes, particularly in improved emissions control. UK development is essentially at the R&D stage, but may be driven towards commercialisation by increasing interest in biomass utilisation.

B. Market Status

Exploitation of specialised wastes has been restricted to traditional opportunistic uses of specific arisings as an adjunct to disposal. Typical examples are furniture manufacturers' use of waste wood and steam-raising from hospital wastes for hospital laundries. Wider uptake as an energy resource is hindered by the discrete nature of the arisings, their generally lower energy density and different combustion characteristics compared with solid fuels. Recently uptake has also been hindered by the trend towards convenience fuels (gas and oil) and away from solid fuels as an industrial energy source – even when this involves a higher fuel price.

As a consequence, utilisation of waste fuels has predominantly been in the hands of the entrepreneur or specialist user, generally operating at small-scale (<<3 MW thermal) often using batch technologies based on simple incineration with heat recovery and little or no automation. The overall performance of such systems, both in terms of conversion efficiency and environmental standards, is generally poor.

In addition, many thousands of small-scale (< 1 tonne/day) incinerators have traditionally been operated in the UK as 'destructors' for commercial wastes by wholesalers, distribution companies and large commercial organisations. These are predominantly batch incinerators, operating on an as-required or single-shift basis without heat recovery and with only rudimentary gas cleaning.

In both cases more stringent environmental legislation introduced subsequent to the Environmental Protection Act has forced the majority of existing plant to close. In the short-term this has resulted in increased diversion of these wastes to landfill. However, environmental pressures on landfilling and other disposal routes may affect this balance significantly in the medium term.

Much of the current activity is focused on clinical waste. Clinical wastes represent only a small fraction of specialised waste arisings, but is one type of waste for which incineration is the preferred means of disposal. The UK has some 900 clinical waste incinerators which treat the bulk of infectious and other wastes arising from hospitals and health care. Few existing plant can achieve, or be modified to achieve, the standards now required under revised legislation. Numerous schemes for replacing obsolete capacity are currently being developed and adopted, often in association with the private sector. This provides a lead market for the introduction of modern, generally larger scale (up to 1 tonne/hour) incineration plant and an incentive for the development of advanced flue gas cleaning systems capable of operating to modern standards at this scale.

Figure 2: Clinical waste for incineration.

The exploitation of specialised wastes overseas is also driven by disposal issues. Niche markets operate where local landfill prices are high or where particular opportunities exist for direct heat sales, combined heat and power or district heating. However, there are also local factors which tend to influence the adoption of incineration for the disposal of municipal wastes. Where the infrastructure permits, specialised wastes tend to enter the municipal incinerator stream, negating the need for new dedicated facilities.

In the UK NFFO has stimulated considerable interest in the exploitation of a wide range of specialised wastes. However, the complexities of securing an adequate waste supply and uncertainties over technological capabilities and environmental legislation have often worked against this interest being converted into financable schemes.

Indications of the current status of exploitation are given in table 1.

Current status	Commercial
UK installed capacity	500-800 MW, predominantly thermal
UK industry	Several established UK equipment suppliers, but technologies require upgrading to modern environmental standards

Table 1: Status of technology.

C. Resource

There is considerable uncertainty over the quantities of individual specialised wastes which could be suitable for energy production. Current best estimates for a variety of waste types are given in table 2.

Waste	Current estimated UK arisings Mtce/year	Energy content Mtonnes/year	Current principal disposal route
Chemical			
Solid/liquid	0.35	0.35	0.16 Mtonnes incinerated either by specialist companies or in-house; few systems with energy recovery. Remainder landfilled with/or without chemical treatment. Disposal costs £2-4,000/tonne
Solvent vapour	0.8	0.8	80% of solvent used lost as low CV vapour; limited number of sites fitted with incinerators. Some liquids recycled for reuse, either in-house or through special companies. Limited incineration of residues; few with energy recovery. Disposal costs £0.1-2 k/tonne
Fragmentiser residues	0.5	0.37	To landfill (£10-40/tonne) on company or private waste contractor landfill sites
Hospital	0.37	0.2	Bulk batch incinerated at hospitals, largely without heat recovery. Disposal costs up to £350/tonne (average £150/tonne)
Meat processing residues/ Animal carcasses	1.75 wet	0.2	Previously processed as protein reinforcement in animal foods; concerns over BSE have resulted in disposal to landfill
Tyres	0.35	0.42	Figures refer to tyres which are surplus to requirements. Landfill and surface stockpile; some incineration with energy recovery. Disposal costs up to £60/tonne (average £40/tonne)
Wood			
Demolition	1	0.6	Bulk to landfill via private contractors. Recycling of demolition rubble opens up opportunities for using timber residue as fuel; disposal costs limited to value of void space (£5 to £10/tonne)
Processing	0.5	0.3	Bulk to landfill via private contractor; disposal costs £15 to £25/tonne. Some on-site use as fuel for space heating

Table 2: Specialised industrial wastes Accessible Resource.

In principle all the resource is Accessible. However, as noted above, the individual arisings are generally quite small, and their disposal dealt with through a multiplicity of isolated private waste disposal contracts. Commercial exploitation depends on aggregating sufficient waste to sustain

Discount Rate	UK Accessible Resource in TWh/year
8%	4.7 TWh/year
15%	4.7 TWh/year

Table 3: Specialised industrial wastes Accessible Resource in the UK at less than 10p/kWh.

an economically-sized plant. For most wastes this would be 5 to 10 tonnes per hour throughput (5 to 10 MW$_e$). Plant at smaller scales would be viable for higher value wastes, such as clinical waste.

Measures introduced under the Environmental Protection Act 1990 to tighten waste management standards will lead to a significant restructuring of the disposal industry and should encourage greater aggregation of arisings. The potential for economic energy production will depend on the delivered cost of the waste at the point of use – a function of both the waste disposal charge levied on the originator and the transportation costs associated with the aggregation process. These costs, and hence the resource potential, are influenced by other factors such as alternative disposal means (mainly landfill), pretreatment required before transport, bulk density in transport and distance between sources of arisings.

Table 3 presents an assessment of the Accessible Resource in the UK, based on the data in figure 3, Section F, for 2005.

Resource Specific Assumptions

Chemical
Combustible wastes from the chemical industry are mainly high calorific value liquids, including, for example, solvent recycling residues. Although not all are strictly 'dry' wastes, all are included here for completeness, giving a current arising of 0.35 M tonnes/year. It is assumed that in the future there will be a greater use of chemical products but that, through waste minimisation and internal reprocessing, the waste arisings will remain constant.

Fragmentation
Car fragmentation currently produces some 0.5 M tonnes/year of combustible waste comprising mainly plastic, rubber (excluding tyres) and natural fibres.

Hospital
Hospital waste includes general industrial waste generated by hospitals plus all clinical waste. In the technical resource it is assumed that all this is available for incineration.

Meat Wastes
It is assumed that all abattoir and other meat waste must be broken down by heating processes. This results in various products and a residue which can be incinerated. The residue, 175,000 tonnes/year, represents 10% of the original tonnage and has a gross calorific value of some 18 GJ/tonnes.

Tyres
The current estimate of arisings is 350,000 tonnes/year.

Wood
Waste wood arisings are estimated at 1.5 M tonnes/year including scrap from furniture manufacture and demolition timber. This is expected to reduce by 10% from present levels in the next few years and then to stay steady.

D. Environmental Aspects

The generic characteristic of wastes is that they are an undesired by-product of other activities. The environmental impact of utilising these residues as a fuel must therefore be placed within the context of other disposal and waste minimisation options.

Stringent environmental legislation for waste combustion processes was introduced under the Environmental Protection Act 1990. Typical stack emissions and other environmental burdens based on modern plant in compliance with these regulations are indicated in table 4. However, the overall environmental impact also depends on other factors such as plant siting and vehicle movements. Each of these requires consideration.

Atmospheric Emissions

Combustion processes result in various emissions to the atmosphere. These include particulate emissions, acidic gases, heavy metals and trace organic materials. The Environmental Protection Act set strict standards for combustion control and flue gas cleaning to prevent or otherwise render harmless emissions. In all circumstances this will require the use of filtration and gas scrubbing equipment and the installation of monitoring systems for process control.

There is particular concern over certain trace emissions, notably dioxins and mercury. The standards now set for these compounds are driving the development of advanced flue gas cleaning technologies and stimulating interest in alternative waste treatment methods, including advanced conversion by gasification and pyrolysis.

These measures are intended to ensure that atmospheric emissions do not have a damaging effect on the environment. A formal evaluation of the impact, however, would normally be required as part of any planning application.

Environmental Burden	Specification (units per kWh$_e$)
Gaseous emissions:	
Carbon dioxide	1.6 kg
Sulphur dioxide	0.5 g[a]
Nitrogen oxides	1.5 g[a]
Carbon monoxide	0.6 g[b]
Volatile organic compounds	0.1 g[b]
Particulates	0.2 g[a]
Heavy metals	0.03 g[a]
Ionising radiation	None
Solid and liquid waste	160 g controlled waste
Thermal emissions	4 kWh[c]
Amenity & Comfort:	
Noise	Subject to planning
Visual intrusion	Some – subject to planning
Use of wilderness area	None
Other	Some – transporting waste to and from sites

Table 4: Environmental burdens associated with specialised industrial waste incineration assuming current trends continue.

Notes   a: Mix of wastes at respective new plant standards
            b: New plant standards, UK
            c: Thermal emission unlikely to be significant for typical plant (some could be recovered as useful heat).

Residues and Gas Cleaning By-Products

The nature of the residues obviously depends on the waste from which they are derived, and on the combustion and gas cleaning technology employed. In some cases the residues can have a market value; for example flyash from tyre burning is rich in zinc oxide. However, flyashes and other gas cleaning residues are increasingly viewed as a special waste because of their potential enrichment with toxic organic and inorganic compounds. These residues, therefore, may require treatment prior to disposal. Grate ashes are generally depleted in such contaminants and may be disposed of more readily to landfill.

Liquid effluents from wet scrubbing or boiler water circuits will generally require some form of treatment prior to disposal. This may be considerable where the effluent is contaminated with heavy metals, particularly mercury and cadmium. There is increasing pressure against any such discharge and, consequently, a move to dry gas cleaning systems.

Visual Impact
Modern large-scale combustion plant can be quite large and often have tall chimneys for the dispersal of stack gases. As they sometimes need to be located near population centres for logistic reasons, visual impact is an important factor in choosing a plant site.

Noise
Vehicle movements, turbines, and general plant machinery can all create a noise nuisance for neighbours of the plant. Careful consideration at the design stage and good housekeeping are the keys to success.

Dust and Odours
Waste deliveries, handling and processing can all create substantial amounts of dust, though this is likely to be a very local problem. Similarly, wastes are frequently biologically active and can emit unpleasant odours during storage or treatment. Good design and housekeeping can minimise these problems. Modern plant often directs air from waste handling processes into the combustion chamber to prevent such emissions.

Vehicle Movements
For some waste to energy plants vehicle movements to and from the site could present a major impact at a local level; this impact therefore needs to be recognised and dealt with in a sensitive way.

Emission savings resulting from electricity generation from specialised wastes are subject to considerable uncertainties, depending amongst other things on the composition of the waste and alternative disposal routes. However, for the purposes of this analysis, it is assumed that 60% of the carbon in the waste is derived from non-fossil sources and can be considered to be in a closed cycle. On this basis a typical plant would displace the emission from average UK generating plant mix as shown in table 5. This estimate is likely to underestimate the savings in emissions from biodegradable wastes since avoided emissions from alternative disposal routes, such as methane escape from landfilling, have been omitted.

Emission	Emission savings (grammes of oxide per kWh)	Annual emission savings per typical 12 MW scheme (tonnes of oxide)
Carbon dioxide	90	8,600
Sulphur dioxide	9.5	910
Nitrogen oxides	1.9	180

Table 5: Emission savings relative to average UK plant mix in 1990 assuming current trends continue.

Table 6 shows the emissions limits required for specialised waste burning plant and the limits that could be expected with further tightening of legislation.

Pollutant	Current trends continue	More environmentally sensitive policies and practices adopted
	Milligrammes per cubic metre	
Particulates	100	10
Carbon monoxide	100	50
Sulphur dioxide	300	40
Nitrogen oxides	250	70
Total hydrocarbons	20	5
Heavy metals	5	1
TCDD 'dioxins'	1.0 ng	0.1 ng

Table 6: Specific emission limits for two different views of the future.

E. Economics

Waste disposal practice inevitably attracts public attention an development of new schemes usually encounters local resistance – a classic case of 'not in my back yard'. Concerns legitimately centre around the likely environmental impact of any disposal site and often involve a public debate about the relative merits of the different disposal options, recycling etc.

Incineration schemes often face criticism. Usually this centres around the problem of emissions, in particular traces of dioxins and heavy metals. Although the levels of pollutants, including dioxins, are now regulated – and in some cases are down to minimum detection levels – the potential toxic effects of these emissions lead to considerable public concern.

For an energy-from-waste scheme to be viable it must constitute the preferred waste disposal option for the particular waste or location. The commercial potential of energy production from wastes in the future will be strongly influenced by developments in environmental legislation and disposal practice. These developments are difficult to predict with confidence, so two different views of the future have been considered. The first assumes that current trends continue and legislation currently proposed takes full effect; the second assumes a more environmentally sensitive future in which environmental controls on waste disposal are more stringent and more emphasis is placed on waste minimisation and recycling.

If current trends continue the following assumptions apply to plant used to generate electricity.

- Conversion relies on the latest steam cycle technology with a conversion efficiency of 25% giving a power yield of 1,215 kWh/tonne.

- Environmental legislation requires adoption of comprehensive emissions abatement technology to meet the standards given in table 6, with the resulting overall plant cost of £2.5 M/MW$_e$.

- Economics are based on plant at the 10 tonnes/hour throughput scale. Cost and operating data are given in table 7.

- All wastes are currently available at a zero gate fee. Improved standards for landfill disposal will increase prices by £10/tonne by 2005 and a further £10/tonne by 2025.

These assumptions are broadly in line with current proposals for new plant. The data is an amalgamation from a number of sources.

If more environmentally sensitive policies and practices are adopted the following assumptions apply to plant used to generate electricity.

- Environmental legislation requires the adoption of sophisticated emissions abatement technology to meet the standards given in table 6. Requirements include catalytic removal of nitrogen oxides and comprehensive ash and residue treatment which consume 20% of the generated power. This leads to overall plant costs of £3.75 M/MW$_e$ installed capacity (£4.7 M/MW$_e$ export).

- Conversion relies on latest steam cycle technology with a conversion efficiency of 25% giving an export power yield of 970 kWh/tonne.

- Economics are based on plant at the 10 tonne/hour throughput scale. Cost and operating data are given in table 8.

- All wastes are currently available at a zero gate fee (i.e. £0/tonne fuel cost). Improved standards for landfill disposal increase prices by £20/tonne by 2005 and a further £20/tonne by 2025.

- Ash and gas cleaning residues are classed as 'special' wastes, requiring separate provisions for landfill disposal. Disposal costs rise to £4.50/tonne fuel by 2005.

Cost and performance data for a 12.2 MW$_e$ plant		
Capital cost	£30.50 M	
Fixed operating costs:		
Maintenance at 2% capital	£610,000/year	
Rates/Insurance @ 1.5 capital	£457,000/year	
Other @ £2/tonne	£160,000/year	
Labour	£300,000/year	
Variable operating costs:		
Residue disposal @ £0.40/tonne fuel	£31,540/year	
Reagent costs @ £5/tonne fuel	£394,200/year	
Project lifetime	20 years	
Construction time	2 years	
Nominal GCV	18.90 GJ/tonne	
Waste throughput	10 tonnes/hour	
Availability factor	90%	
Conversion efficiency (fuel to electricity)	25%	1315 kWh/tonne
System load	7.5%	100 kWh/tonne
Power yield		1215 kWh/tonne

Table 7: Plant costs and performance data assuming current trends continue. Costs are for all new equipment and associated civils.

Cost and performance data for a 9.7 MW$_e$ plant		
Capital cost	(£M)£45.75 M	
Fixed operating costs:		
Maintenance at 2% capital	£915,000/year	
Rates/Insurance @ 1.5 capital	£686,000/year	
Other @ £2/t	£160,000/year	
Labour costs	£300,000/year	
Variable operating costs:		
Residue disposal @ £4.50/tonne fuel	£354,800/year	
Reagent costs @ £15/tonne fuel	£1,183,000/year	
Project lifetime	20 years	
Construction time	2 years	
Nominal GCV	18.90 GJ/tonne	
Waste throughput	10 tonnes/hour	
Availability factor	90%	
Conversion efficiency (fuel to electricity)	25%	1315 kWh/tonne
System load	26%	343 kWh/tonne
Power yield		972 kWh/tonne

Table 8: Plant costs and performance data assuming more environmentally sensitive policies and practices are adopted. Costs are for all new equipment and associated civils.

F. Resource-Cost Curves

A number of common assumptions have been made in establishing the Maximum Practicable Resource. The starting point is the current waste arisings, described in table 2. These wastes have a range of characteristics and attract a range of disposal fees. It is assumed here that these characteristics and fees can be represented by weighted mean values. These values, however, change with both time and scenario in line with the above assumptions and with the population, transport and industry assumptions given in the scenario documents. The resulting resource data are displayed in figure 3.

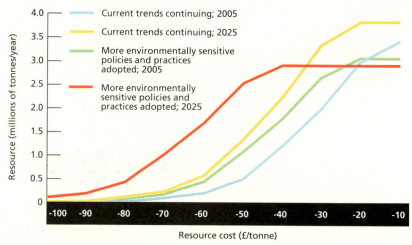

Figure 3: Specialised industrial wastes resource.

For most dry wastes the aggregation of sufficient quantities for viable schemes is constrained by the current disposal infrastructure. It is assumed that, through promotion and information dissemination, larger and more robust organisations will be encouraged to utilise wastes for electricity generation and that the infrastructure will be developed accordingly. Under such assumptions the entire resource can be accessed, at an appropriate cost.

Resource-cost curves have been generated by combining the economic data presented in Section E with the resource data presented in figure 3. Discounted cash flow analyses over a 20 year plant life have been carried out at 8% and 15% discount rates. These rates may be considered low for capital-intensive schemes such as energy-from-waste plant. However, developers are generally in competition with alternative low-cost disposal, mainly landfill, and therefore need to accept a lower return against a longterm waste disposal contract. The exception is with clinical waste incineration plant for which the pace of regulatory change is expected to require substantial modification to plant on a four to eight year cycle. This refinement is not included in the modelling exercise reported here.

The graphs presented below are for the Maximum Practicable Resource which takes account of changes in the available resource with both time and view of the future.

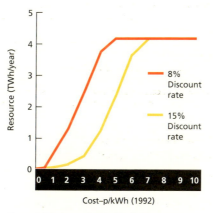

Figure 4: Maximum Practicable Resource-cost curves for specialised industrial wastes in 2005 assuming current trends continue.

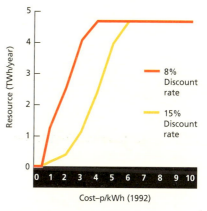

Figure 5: Maximum Practicable Resource-cost curves for specialised industrial wastes in 2025 assuming current trends continue.

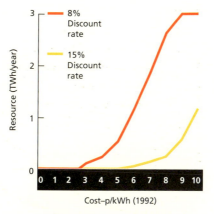

Figure 6: Maximum Practicable Resource-cost curves for specialised industrial wastes in 2005 assuming more environmentally sensitive policies and practices are adopted.

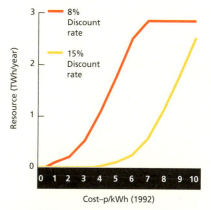

Figure 7: Maximum Practicable Resource-cost curves for specialised industrial wastes in 2025 assuming more environmentally sensitive policies and practices are adopted.

The variation in the cost of electricity is due to the variation in the assumed gate fees for waste disposal. However, the gate fee for any particular scheme will be subject to competitive pressures and in cases where the calculation results in electricity costs below its market value it would be more realistic to do the calculation the other way round, taking the electricity price as given and calculating the gate fee required to cover costs.

G. Constraints and Opportunities

Technical/System

The technology for energy-from-waste schemes must adapt to the new environmental requirements. Confidence needs to be raised in the technology. This will require further evidence of successful operation. Successful NFFO projects could provide such examples.

In practice, completing the development and demonstration stages in the evolution of these technologies may prove difficult. The plant has to be close to full scale if meaningful lessons are to be learnt. The companies currently involved in these developments in the UK often do not have the resources to finance extended development work at larger scale. This may lead to a general reliance on overseas technologies. However, lack of appropriate experience can lead to premature projects entering the commercial market without adequate contingency. When technical difficulties are encountered with such projects, both inadequate financial provision and lack of contractual flexibility may prevent resolution of the problems.

Institutional

Industry is under increasing regulatory and public pressure to control waste arisings and to manage more stringently the disposal of residues which cannot be recycled or re-used. This is resulting in a reconsideration of the traditional reliance on low cost waste disposal contractors and low cost landfill.

Energy recovery is increasingly viewed as an option to be considered as part of responsible waste management. However, in order to be cost-effective, the systems must provide disposal at a competitive cost and energy at a cost competitive with other sources. Consequently, there are interlocking issues beyond those of the technology itself which will determine the viability of energy-from-waste schemes. Lack of experience in dealing with these issues, and the evolving regulatory framework for environmental protection, present significant barriers to the uptake of schemes. Duty of care requirements, however, will push waste disposal into the hands of a smaller number of larger, better informed, disposal companies. This may provide the opportunity to aggregate the arisings into quantities required for commercially viable schemes (5,000 to 10,000 tonnes/year).

NFFO has also stimulated the interest of companies with experience of both the technical and commercial facets of energy production, for example heat service companies, the privatised electricity production companies and the RECS, as well as some entrepreneurs who recognise that this area may provide niche business opportunities in the future.

Financing projects in this area poses some problems. There has been a history of difficult and unsuccessful projects; this has given the area a poor reputation. Projects tend to be organisationally complex, with the economics dependent on many parameters. In some cases these problems have been circumvented by the establishment of joint ventures, but this pattern is not yet well established.

H. Prospects Modelling Results

Resource-cost curves are presented in figures 4-7. These indicate a considerable sensitivity to future environmental policy, as it affects not only the scale of waste arisings but also disposal costs and environmental protection costs reflected in the capital cost of plant. Under CSS assumptions uptake of up to 1.5 TWh/year by 2005 could be possible, rising by a further 2.0 TWh/year by 2025 as a result of increased costs of alternative disposal. However, the high capital cost of plant makes such predictions highly sensitive to discount rates, and uptake may be reduced to below 1 TWh/year by 2025 if the higher discount rate is applied.

These trends are exacerbated in more environmentally sensitive scenarios, particularly as a result of higher capital investment in environmental protection measures to meet tighter environmental standards. Uptake by 2025 may not exceed 1 TWh/year even at the most favourable discount rates under electricity price and environmental assumptions for the HECa scenario.

These modelling results underline the uncertain contribution of incineration to future disposal practice for industrial wastes. However, as noted above, the greater private sector involvement in municipal waste disposal is anticipated to result in an increased assimilation of industrial wastes into the municipal waste stream, particularly where destined for resource recovery. This is increasingly evident in overseas practice. Uptake by this route is subject to the same factors and technological choices which determine the assimilation of general industrial wastes, as discussed in the municipal waste module.

It is anticipated that niche markets will remain for the uptake of specific arisings under locally favourable conditions or where there are particular constraints on alternative disposal routes. Clinical waste has been cited as the principal example for such exploitation, but incineration may also provide a desirable option for scrap tyres, fragmentiser residue, certain hazardous chemical wastes and contaminated wood. Local circumstances could give rise to significantly different resource-cost relationships to those presented here, dependent primarily on the waste disposal credit available. It is assumed here that industrial wastes have a mean energy yield of 1,215 kWh/tonne so that an additional disposal fee of £12/tonne would provide additional income equivalent to 1p/kWh. More significantly, local opportunities may exist for CHP or direct heat sales which would radically alter local economics. The long term potential and prospects for industrial waste utilisation may depend significantly on such niche opportunities.

II. Programme Review

A. United Kingdom

Incineration has been viewed as an opportunistic means of waste disposal. Little systematic development has been undertaken by manufacturers to improve the basic technology used over past decades. Current plant principally relies on small-scale batch-fed systems operating on a single shift or as-required basis.

Government programmes oriented towards waste disposal have traditionally been maintained by the Department of the Environment and, to a lesser extent, by the Department of Trade and Industry. Incineration has not enjoyed a prominent position due to the small-scale and fragmented nature of current UK practice. The work undertaken has concentrated largely on controlling combustion conditions to minimise stack emissions. Little has been done, outside the former Department of Energy Programmes, on the potential for waste as an energy resource. The introduction of NFFO, however, has accelerated commercial interest in energy from waste and has stimulated a number of major schemes. It is expected that these schemes will act as a further spur to UK development.

The Government's Programme

Since publication of Energy Paper 55, considerable strides have been made under the former Department of Energy Renewable Energy and Energy Efficiency Programmes in the utilisation of waste fuels. Much of this work has been geared to developing and demonstrating appropriate combustion technologies and the subsequent dissemination of the results through publications, workshops and seminars. Technology development has largely been undertaken in collaboration with individual entrepreneurs at a relatively small-scale, chiefly for steam raising or CHP.

Latterly, NFFO has attracted several larger scale projects and the Programme has moved increasingly towards project monitoring. The current Programme plan aims to build on the previously sponsored work on specialised wastes by collaborating with industry to bring forward a range of R,D&D studies geared to assessing the resource, developing and promoting appropriate technologies and understanding and helping to overcome the chief non-technical barriers. In collaboration with other government departments, the Programme aims to support commercialisation initiatives with the objective of improving public and industrial perception of energy recovery from these wastes.

A watching brief is being maintained on a wide range of other industrial and commercial wastes. Although many schemes have been discussed, uncertainties over the capabilities of modern incineration and gas cleaning plant and uncertainties over future environmental standards are acting as major barriers to the commercial realisation of these schemes. Projects undertaken under the Pollution Abatement programme are providing much of the missing information and, consequently, are generating considerable industrial interest.

B. Other Countries

The utilisation of waste fuels overseas is very much constrained by the factors which influence the UK situation. Waste fuels are therefore utilised on an opportunistic basis where the economics are favourable against other means of disposal. Generally, this has been at small-scale. However, where a suitable infrastructure for waste disposal exists, larger scale schemes have been developed, often in association with municipal waste incineration. Consequently, examples of most types of technology can be readily identified. However, generally these have also been affected by recent regulatory changes towards more stringent emissions control. Few, therefore, represent reference facilities upon which UK developers can draw. As standards tighten further, the lack of suitable reference plant may become a universal impediment.

The potential offered by wastes through advanced combustion technologies, which increase the energy conversion efficiency, has been widely recognised. Considerable R,D&D is therefore underway. Generally, this is geared towards specific feedstocks of particular interest to individual organisations or Governments. Much of this work is accessible through the information gathering and dissemination activities of the International Energy Agency in which the UK is an active participant. The European Commission, the World Bank and other funding agencies are also actively engaged in technology development, especially for applications in developing countries where conversion of local biomass feedstocks to electricity offers significant potential.

Agricultural and Forestry Wastes

I. Technology Review

A. Technology Status

Agricultural and forestry wastes fall into two main groups – dry combustible wastes such as forestry wastes and straw, and wet wastes such as green agricultural crop wastes (e.g. root vegetable tops) and farm slurry. The first group can be combusted, gasified or pyrolysed to produce heat and/or power. The second group are best used to produce methane-rich biogas through the process of anaerobic digestion.

Dry Agricultural & Forestry wastes

Straw

Straw is available from cereal and other 'combinable' crops such as oilseeds. It is produced seasonally and is localised, with highest production centred in East Anglia. Straw is produced at cereal harvest and usually left on the ground after the passage of the combine harvester through the crop. It must then be recovered and baled in a second pass operation. As straw is a low density material, transport and storage costs can be high. This has led to the adoption of large high density Hesston bales, with baling operations increasingly being carried out by contractors. Nevertheless, farmer owned and operated machinery still makes a significant contribution.

Straw is a relatively low quality fuel, with an energy content around 18 GJ/dry tonne, which is independent of straw type. It is harvested with a low moisture content of around 15%. At this moisture content the calorific value of the straw is around 15 GJ/tonne. The major problem associated with the use of cereal straw as a fuel is the low temperature at which ash fusion occurs. This can lead to boiler slagging and fouling and necessitates careful design and good combustion control. This is less of a problem with oilseed straws, which produce ash with a higher melting point. As with other biomass fuels, co-firing with coal is a feasible option to give benefits of higher conversion efficiencies in a larger plant, shared costs for infrastructure, capital, labour, etc. and reduced emissions from the fossil fuel. However, co-firing with coal is likely to increase maintenance costs of the fossil fuel plant and the risks of boiler slagging and fouling. Also, by co-firing with a fossil fuel the ash from a biomass plant creates a disposal problem.

The use of straw as a fuel is commonplace at the small scale (up to 100 kW_{th}), using whole bale burners. These operate as batch-fed devices, mainly for on-farm domestic space heating. In Denmark, there are examples of the larger scale use of straw for district heating schemes (at around 10 MW_{th}) and power generation using CHP at around 30 MW_{th}.

Plant is available in a wide range of sizes. The upper limit of this range is determined by the mechanics of feeding the fuel into the combustor without densifying it or introducing other technical refinements. Some of the available plants use 'cigar-burner' technology, where whole bales are fed into the combustion chamber in such a way that they burn from one end. Other systems employ automated batch feeders using whole bales. Conventional stepped grate technology is also suitable for straw, and modified wood burning systems are widely used. Suspension combustion devices such as cyclone burners can also be used, as can suspension firing over a grate of either coal or wood fuel. All these plants have a life-expectancy similar to that of the coal combustion plants from which they are

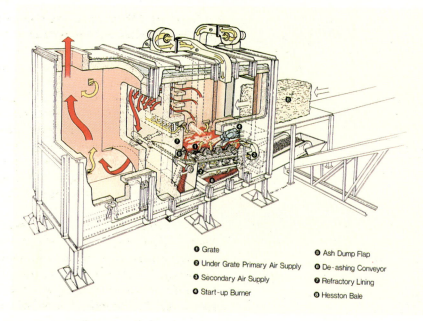

○ Grate   ○ Ash Dump Flap

○ Under Grate Primary Air Supply   ○ De-ashing Conveyor

○ Secondary Air Supply   ○ Refractory Lining

○ Start-up Burner   ○ Hesston Bale

Figure 1: Cutaway of a whole-bale straw burner.

derived – something in excess of 20 years. After this time the plants will incur similar decommissioning charges to fossil fuel plants with the possibility that the scrap value will cover these costs. Alternatively, boiler replacement and plant refitting can be considered.

It is difficult to describe a typical scheme as these will vary with location. For instance, in East Anglia, close to the area of highest straw density, a few large (30 MW$_e$) schemes may be feasible. In other areas the amount of straw may only support schemes in the 10 MW$_e$ range. In planning these projects issues such as the availability of good road access, cooling water and a grid connection point will be of paramount concern. Some conflict may arise over these siting issues in the east of England where the flat terrain will render plant highly visible. These potential problems have yet to be quantified, but they will be addressed when the first demonstration projects are developed.

Due to the seasonal harvesting patterns and low density of the fuel, extensive storage facilities must be available. These can be covered stores located at the point of use, but will more likely consist of on-field stores in which straw is covered with tarpaulins.

Forestry Wastes

Wood is man's oldest fuel and is still the major source of energy throughout the developing world. The UK is one of the most sparsely wooded countries in Europe, with only about 10% of its land area covered with trees. The resource is also spread disproportionately around the country with about half located in Scotland and Wales.

Forestry management is designed to give a timber crop. Production cycles for timber vary from 40 to 80 years for conifers to 80 to 120 years for hardwoods. The highest timber values result from producing straight, knot-free stems. This is achieved by planting trees close to each other to stimulate rapid, straight growth, and then thinning them at intervals to maintain the maximum growth rate. Harvesting and thinning operations continue throughout the year. As the value of the tree is in the stem, typically only that part of the tree is removed at final harvest (clearfell) in this country, the unwanted branches and tree tops being removed from the tree on felling and left in the forest as residues (or brash). Up to 50% of the above ground biomass can be discarded in these operations. This brash presents a barrier to restocking.

The UK forest harvesting industry is increasingly mechanising to reduce costs. This is in line with the practice in other countries, but the UK lags behind them by some years. Modern harvesters are made up of a base unit and a hydraulically driven harvesting head mounted on a mechanical arm. The head operates by gripping the tree, cutting through the stem with a saw, turning the cut stem through 90 degrees, driving the stem through delimbing knives mounted on the head, and cutting the stem to the required lengths. In the UK thinning is still commonly carried out manually with a chain-saw. In other countries mechanical methods are being adopted.

Most forest owners do not harvest their own trees but employ a contractor. The forest industry is composed of a number of harvesting contractors, ranging in size from single machine owner/operators to large national companies. Trees are often sold standing, with the owner negotiating the sale either with a merchant or with the harvesting contractor who then sells the timber. The residues left on site usually remain the property of the woodland owner who may employ a second contractor to clear the site and replant the trees. For example, the Forestry Enterprise (recently created as part of the Forestry Commission) currently excludes the brash material when selling standing trees and offers a second contract for the collection and removal of the brash.

Figure 2: Forest residues: integrated harvest in progress.

The provision of fuel wood from existing forestry operations will require a second pass residue harvesting operation, with the brash being either removed from the forest using a modified tractor (a forwarder) for chipping at the roadside (the landing), or picked up from the floor and directly chipped using a specialist terrain chipper. The former option is less costly as a contractor only needs to purchase a suitable chipper if he does not already have one available. Chipping at the landing can also be more efficient because the relatively stable environment allows greater chipper throughput. However, as a second pass operation the cost of fuel wood harvesting from residues is high. A better alternative is to use a one pass integrated harvesting approach where the tree is harvested and removed from the wood whole for product separation and processing at the landing. This method requires increased mechanisation and, while the overall harvesting efficiency is also increased leading to lower stem wood production costs, investments of this magnitude (around £300,000 per harvesting operation) will only become economically viable when a fuel wood market is firmly established.

Wood is a low quality fuel, having a calorific value around 19 GJ/dry tonne irrespective of tree species. When harvested, wood has a moisture content in the order of 55% by weight. The calorific value at this moisture content falls to around 10 GJ/tonne. Wood has a low sulphur content and gives only 1% ash. Wood fuel is used domestically or industrially in a large number of countries. For domestic use, the wood is usually burnt as logs in stoves or on open fires. For industrial use it is chipped prior to combustion in fully automated specialist plant. Plant sizes thus vary widely.

Figure 3: Drying kiln fired by forest residues.

Most wood combustion plant from 0.1 MW_{th} upwards are based on coal combustion sloping grate technology, with the addition of a refractory lining and the provision of secondary air. Wood can also be combusted using fluidised bed technology. Both these approaches allow wood with moisture contents as high as 55% by weight to be burnt efficiently. As these plant are based on coal technology, the plant life can be expected to be similar to that for coal fired plant - that is, around 20 to 25 years. After this time the plant can be decommissioned (at a cost similar to that for coal plant), or new plant can be installed on the site as required to continue electricity generation.

Typically, the size of a project will be limited by the availability of forestry wastes in the region. Thus 5 to 15 MW_e will be a typical size range. However, it is possible that the fuel resource can be increased by growing arable coppice as an alternative farm crop. In this case plant as large as 30 MW_e may become feasible. These are likely be located in remote, typically wooded areas which can provide adequate screening. However, the choice of site will depend on many factors such as the availability of adequate roads, cooling water, grid connections etc. Thus the overall visual impact for planning purposes will be site specific, and can only be addressed when the first demonstration projects are being planned.

Since wood fuel is produced from forestry operations throughout the year, the only storage facility necessary is a buffer store. If properly constructed, these stores can also be used to dry the wood to some degree. Incorrectly constructed stores can result in degradation of the chips due to microbial action. This can lead to significant dry matter loss.

The gasification of wood is a well known technology, even though it is only just entering the commercial exploitation phase. Gasification requires a dry feed stock; this can be achieved by installing a drying loop in the fuel supply system.

Wet Agricultural Wastes

Animal Slurries
Animal slurries are derived from two major sources, cattle and pigs. These slurries are many times more polluting than human sewage and when not correctly managed can cause serious environmental damage, particularly by adulterating water courses and producing odours. Fines for this kind of

pollution can now be as high as £20,000 per offence. This problem is less intense, and also less manageable, when animals are reared in the open. In this case the slurry is not collectable and must be controlled by the correct management of drains and the installation of suitable ditches.

The problem becomes acute in the vicinity of milking parlours and during the over wintering of animals indoors. In these cases costly slurry handling and management systems are required and in the UK these are installed on individual farms. They usually consist of floor scrapers to move the slurry into channels from which it is pumped into a storage tank or lagoon, sized to accept all the slurry produced during the winter as well as water run-off. Sophisticated systems also use a separator to remove the fibrous portion of the slurry for sale as a compost base, while the more basic systems allow natural settlement. The slurry is ultimately returned to the land during the summer.

The feedstocks are composed of livestock slurry (faeces plus urine), dilution water from open yards, bedding materials (straw, sawdust, peat and waste paper), the concentrated effluents from ensiling crops (grass and maize), and surplus vegetable crops and residues. These materials can be digested in a similar way to sewage sludges, but they need to be diluted and balanced by the addition of organic matter to provide optimum digestion characteristics and satisfactory gas yields. Some of the feedstocks have high nitrogen contents while silage effluent has a high volatile fatty acid content and low pH. Livestock slurries are often diluted with yard and roof water to such an extent (less than 2% dry matter) that economic gas production is no longer feasible without the addition of more dry matter such as poultry manure.

There are three basic options for managing livestock slurry; these are set out in table 1. At present most farms operate Option 3, which consists of collection from the buildings and storage for not less than four months, then application to agricultural land. A few hundred farms separate the slurry before storage. The separated fibre is either composted and sold for gardening or used on the farm.

Options 1 and 2 involve biological treatment by anaerobic digestion or by aeration. Digestion is carried out on farms where there is an opportunity for using the biogas produced for heating or for the generation of electrical power. Aeration is generally carried out on farms which have an odour nuisance problem; no energy component is produced in this case.

Anaerobic digestion is the bacterial fermentation of waste organic material producing biogas made up of 65% methane and 35% carbon dioxide – the calorific value of this biogas is about 25 MJ/cu.m. Typically, between 40% and 60% of the organic matter present is converted to biogas; the remainder consists of a stabilised residue with some value as a soil conditioner.

Option 1	Anaerobic digestion	Fibre separation	Compost production (from solids)	Spread liquid on the land
Option 2		Fibre separation	Compost production (from solids) after aerobic digestion	Spread liquid on the land
Option 3			Whole slurry storage	Spread whole slurry on the land

Table 1: Slurry handling options.

Cattle	20 days
Pigs	10 to 12 days
Poultry	20 to 25 days

Table 2: Typical retention times for optimum economic gas production.

Figure 4: Farm waste digestion plant.

Farm digesters are similar in design and operation to those used for sewage sludges, except that the tanks are cheaper, being made in situ of reinforced concrete, fibreglass or vitreous enamelled steel. The digesters can be either above or below ground level insulated tanks, with internal heat exchangers to maintain temperatures typically within the mesophilic range of 30°C to 35°C. The tank content is usually mixed by re-circulating biogas. The digesters operate mainly as plug flow systems.

A recent development has been to adapt digesters for higher dry matter wastes such as farmyard manure with substantial quantities of straw. These produce a high yield of fibre in the digested material which can be processed further by composting to produce a high grade peat substitute for horticulture and for local authority landscaping in towns and cities. Compost sales have a substantial impact on the overall costs of the system and improve the profitability of the enterprise.

The technology of anaerobic digestion is now well developed and a range of digesters from 70 cu.m capacity to 5,000 cu.m are commercially available. The size and type depends on the manufacturer and the quantity and type of the material to be digested.

The methane produced is used either in gas engines for electricity and combined heat and power generation, or in gas boilers modified to accept a lower calorific value gas. These boilers can be used for process or space heating. The size of the on-farm units in a typical UK application will be in the 100 kW$_e$ range. They will be built into the existing farm infrastructure so that disturbance and visual impact will be kept to a minimum. Centralised anaerobic digestion plants are likely to prove the most viable option following the Danish experience; in which case, the plants will be of the order of 1 MW$_e$.

Green Agricultural Wastes
These wastes mainly arise from the processing of root vegetables and sugar beet, as well as from the run-off from the ensiling processes, described above. The wastes are produced seasonally and can present the farmer with a disposal problem. Currently, they are either ploughed in green, allowed to rot prior to ploughing in, or merely left to decompose. A further option is to use the wastes as animal feed, either as they are produced or after ensiling.

The use of these wastes as a fuel has not been seriously considered. However, they can be co-digested with slurry to produce methane, especially if a centralised digester facility is employed. Here, the addition of green wastes will increase both the amount of methane produced and the fibre content, and therefore the value, of the final compost.

B. Market Status   ***Dry Agricultural & Forestry Wastes***

Straw
About half the cereal straw produced annually is used for agricultural purposes, mainly as animal bedding. This market is typically localised away from the cereal areas of the country, so areas with extensive livestock farming and low cereal production need to import straw. Cereal straw is also used to protect some crops, such as roots, against frost and to keep others, such as strawberries, off the ground. Oilseed crops are not suitable for either of these uses.

Other potential industrial uses for straw are for paper making, or in the production of industrial fibre, including constructional board manufacture. Here it is the internode (i.e. the stem between the points at which the leaves grow) which is of most value and mechanisms to extract this portion of the straw are being developed.

In the past, when markets were not available for the straw, it was burnt in the field along with the stubble. This practice was banned from the 1992 harvest onwards, leaving incorporation, (ploughing in) as the major alternative. This practice does have some value as straight nutrient replacement on some soils, but deep cultivation is often required and this is a highly energy-intensive process. In addition, there is some doubt as to the long term viability of this disposal option for particular soils.

The only country using straw for power generation is Denmark, where taxes on imported fossil fuels make the conversion of agricultural wastes such as straw economically attractive. The country now boasts some 300 MW_{th} of installed capacity.

With support through the NFFO the technology, which already exists, could be demonstrated in this country and the infrastructure of a fuel supply created. Under these circumstances, farmers would be willing to invest in appropriate mechanisation to reduce the delivered price of the fuel. In addition, British boiler manufactures would be encouraged to develop new, cheaper plant to use straw fuel. The resulting competition will have the effect of reducing the costs of electricity generation.

Forestry Wastes
The markets for the products of forestry thinning operations (small roundwood) are varied. The main use is for paper and particle boards, but traditionally other markets such as pit-props and fencing have been significant, though these are now in decline. However, all these markets tend to be volatile and subject to interference from cheap imports from Europe and Canada. Clearfell residues have a less well defined outlet. Given the nature of the resource, it is common for the wood to be left on the forest floor to rot. Where this material is recovered, it is comminuted (usually by chipping) to decrease the handling costs. Wood chips can be marketed for horticultural mulch or equestrian flooring. Where a white (bark-free) chip can be screened out, it can be used in the manufacture of paper or particle board.

Electricity generation from wood fuel is restricted in the UK to locations where sawmill or paper making residues are co-fired with fossil fuels in existing combined heat and power plants. The extent to which this practice is currently being employed is unknown, but it is unlikely to yield more than a few MW. This is in complete contrast to the situation in North America, where generation from wood is commonplace and where there is some 6,000 MW_e of installed capacity. On the other hand, in Scandinavia, power generation is not so common. Instead wood fuel is routinely used in district heating plant.

The NFFO will allow some of the technologies for producing energy from forestry wastes to be demonstrated in the UK. It will also lead to the development of the required fuel supply infrastructure, and provide opportunities for the forest products industry to adopt appropriate mechanisation to reduce the delivered price of the fuel. In addition, by creating a market for wood fuel, it will encourage British boilermakers to develop appropriate plant. Additional competition in this sector will help to bring down plant cost.

Wet Agricultural Wastes
Currently, there is no market for animal slurries or green agricultural wastes.

The main barrier to large scale expansion on farms has been the initial cost of investing in a digestion system when farm incomes have been falling over the past two or three years. There has been relatively little pressure to invest in pollution abatement systems until very recently.

In other countries, such as Holland and Denmark, the slurry handling problem is removed from the farm. There, groups of farmers have co-operated to set up central anaerobic digestion facilities to provide an off-farm diversification. The slurry is removed by tanker to the plant where large scale digestion is carried out. This allows both the blending of wastes in the tank and control of the digestion process. In this way higher temperature thermophilic processes can be operated. These offer several advantages, including higher methane production, faster throughput, better pathogen and virus 'kill' and the prospect of compost production to a consistent standard. A further advantage is that plant of this nature can handle other biological wastes such as food processing wastes or even human sewage, both of which will attract a gate fee. Currently about 10 MW_e of capacity is installed or planned in Denmark and Holland. This compares with 1 MW_{th} in the UK.

There are a number of manufacturers of anaerobic digestion plant in this country and abroad. The working lifetime of the digestion and gas storage tanks is, in theory, limited only by the structural integrity of the tanks, which at a conservative estimate should exceed 25 years. Other items such as pumps will require routine replacement.

Through the NFFO, it should be possible to demonstrate the operation of a large off-farm facility of the type operating overseas. The economic advantages of this route over the small on-farm facility will then become apparent and this should lead to further economic exploitation of the available resource. An additional benefit of the NFFO is that it will lead to an earlier deployment of the large resource offered by energy crops.

	Straw	Forestry Wastes	Animal Slurries	Green Agricultural Wastes
UK installed capacity	None for power generation, approximately 300 MW_{th} for space heating	2 MW_e as CHP in the wood processing industries (estimate)	1 MW_e	None
World installed capacity	50 MW_e	10,000 MW_e	Unknown	Unknown
UK industry	Farming industry well placed to supply but will need to use more mechanisation. UK boilermakers currently developing suitable plant at the pilot scale	Forest products need to mechanise in order to supply fuel economically. UK boilermakers do not currently manufacture suitable plant	UK farming well placed to exploit the resource. Some UK suppliers of slurry handling, digestion and biogas utilisation plant	As for animal slurries

Table 3: Status of the technologies.

C. Resource.   *Dry Agricultural & Forestry Wastes*

Forestry Wastes

The area currently under woodland in the UK is estimated at between 2 and 2.7 million hectares (hectare (ha): 10,000 sq.m). This figure is rising by around 20,000 to 25,000 ha/year. By 2010 the area of woodland may rise to some 3.2 million ha, provided there are sufficient incentives to promote large scale planting. Around half the existing resource is in private ownership; the remainder is owned by the Forest Enterprise which has recently been formed from the Forestry Commission.

In 1989 the Forestry Commission estimated that some 7.4 million tonnes of wood was harvested in the UK. This represents 12% of our annual demand for wood. Of the 7.4 million tonnes, only 14% (1.06 million tonnes) was used as a fuel, and this almost exclusively as firewood.

As a result of the forest management and harvesting practices currently in operation (as described in section A), as much as 4 million tonnes per year of residue material is left in the forest post-harvest. This resource is dispersed throughout the country, generally in remote locations so that the fuel is produced at sites well away from centres of population. This means that the most attractive way of using this energy would be to produce electricity for distribution to centres of demand. As much as 2.8 TWh/year could be provided in this way.

Whilst this is the Accessible Resource, it should be considered as a conservative estimate. This is because the small blocks of private woodland (especially farm woodlands) are difficult to survey but are good potential sources of wood fuel as often the woodland has become derelict and contains only low grade timber for which there is little demand. As a result the cost of rehabilitating these woodlands is prohibitively high, but the sale of wood into an energy market may provide sufficient income to make this a viable option for the woodland owner. Because this wood fuel is in small dispersed blocks and only transiently available it is unlikely to be used for power generation. However, it may be used to supply small heat markets, or more importantly as a precursor to energy crops where this resource can be mobilised to achieve early market development.

Straw

In 1990, straw was being produced at the rate of 7 million tonnes/year. This level of production will ultimately be reduced by reform of the Common Agriculture Policy (CAP) by the European Union. A main element of this policy is Set-Aside, which is a measure to reduce the overproduction of food by effectively subsidising farmers to leave fallow parts of their land. In 1993, 15% of arable land was eligible for Set-Aside. This equates to some 630,000 ha. At a typical straw yield of 3 tonnes/ha, the size of the resource may therefore be reduced by up to 1.89 million tonnes at the 1993 harvest. However, this assumes that all the Set-Aside land was formerly used for cereal production and that the Set-Aside measures do indeed lead to a reduction in output. There is some evidence that introducing a fallow year into the cropping cycle actually leads to increased yields in the following years.

The straw resource, like the wood resource, is produced in the countryside away from major areas of population. Thus, as with wood, electricity generation is the best option for this fuel. If all the straw produced (7 million tonnes/year) were used for this purpose, 7.29 TWh/year would be generated. This is the Accessible Resource.

Wet Agricultural Wastes

The total Accessible Resource available from farm slurries is sufficient to generate 2.86 TWh/year. Estimates of the potential resource from green agricultural wastes are more difficult to make as there are no up-to-date figures for these. However, a study carried out in 1985 indicated that 1 TWh/year could be generated from these wastes alone. This is the Accessible Resource size. However, in practice these wastes are only likely to be used where they are co-digested with animal slurry (though even here it is likely that only about 25% of the feedstock can be made up of green wastes). This is a major constraint on the use of these wastes as a feedstock for energy production because they are generally produced in arable, not dairy areas. However, they could be transported to digesters if this was an economically attractive option. As a result of all these constraints, the Maximum Practicable Resource is estimated at 25% of the Accessible Resource.

The forestry waste, straw and wet agricultural waste resources measured as annual tonnages are assumed to remain the same for 2005 and 2025. However, the resources in terms of electricity generated will increase as new technologies with higher conversion efficiencies become available. For straw, combustion may be replaced by thermal conversion increasing the efficiency from 25% to 35%. For forestry waste, gasification technology is assumed to improve from 35% to 45% efficiency. The resultant estimates of Accessible Resource are shown in table 4.

	Accessible Resource	
Year	2005	2025
Straw	7.29	10.2
Forestry wastes	3.89	5.0
Animal slurries	2.86	2.86
Green farm wastes	1.0	1.0

Table 4: Accessible Resource (TWh/year) at less than 10p/kWh for both 8% and 15% discount rates.

D. Environmental Aspects

Dry Agricultural & Forestry Wastes

In general, the dry combustible fuels will be used as a direct replacement for fossil fuels in power generation or for industrial heating. This replacement of fossil fuels with carbon dioxide neutral fuels, which are also low in sulphur, will reduce emissions.

However, if wastes are to be used as fuels they will require transport from the site of production to the site of use. This will be an inefficient and costly business as the wastes have low bulk densities so that transport vehicles cannot be loaded to their maximum axle weights. In addition these wastes have a relatively low energy content per tonne when compared with solid or liquid fossil fuels. This reduces the overall energy balance associated with using these wastes as fuels and increases the emissions associated with their use. This situation can be improved to some extent by siting the utilisation plant close to the point at which the fuels are produced.

Forestry Wastes

There are several points to bear in mind in any consideration of the use of forestry wastes as fuel.

Firstly, forestry sites are more amenable to early restocking if cleared of residues. Cleared sites are also less likely to develop disease problems from the contaminating of new trees by brash. Moreover, the sites are more visually attractive when cleared of discarded residues. However, the cropping and removal of the whole tree from forest sites is a cause for concern in that nutrients are also being removed. Moreover, after a site has been cleared there is more water run-off which generally leads to soil erosion. According to published evidence, this results in more nutrients being leached from the site than are lost from brash removal. This condition may be further exacerbated if no brash is present. It must be said, however, that there is considerable disagreement on this matter. Some published reports find that whole tree harvesting techniques have no major impact on nutrient removal.

Another major concern is soil compaction. Currently, the residue material provides a mat over which machinery can operate without serious soil compaction damage. If this material is removed, compaction damage may well occur. This problem can be minimised by using low pressure 'flotation' tyres or tracks on the vehicles, or by managing the site so that a residue mat is created on areas of high traffic movement.

The ash produced from combustion of the fuel (representing only about 1% of the fuel) presents no disposal problems as it is not contaminated and has value as a fertiliser. Indeed in Scandinavia it is common for the ash to be returned to the forest as a means of nutrient recycling.

Wood burning electricity generation plant will most appropriately be sited off the farm or forest in a location offering good access for fuel supply movements as well as access to grid connections and water supplies.

Figure 5: Fibropower Ltd's power station at Eye in Norfolk, fuelled by chicken litter combined with other dry agricultural waste.

Wood fuel buffer stores are most likely to be sited in remote forest locations close to the site of production. These are unlikely to cause significant visual disturbance as forestry wastes are harvested year-round and therefore require less storage space than seasonally harvested energy crops.

Straw

From the 1992 harvest, straw burning in the field is banned. This may present some environmental problems associated with straw incorporation which is the most likely route for straw disposal in the absence of an energy market for this fuel. The long term effects of routine ploughing-in are not yet known.

As with wood fuel, the ash left after straw fuel burning has value as a fertiliser and can be returned to the soil.

Straw burning electricity generation plant will most appropriately be sited off the farm, in a location offering good access for fuel supply movements and access to grid connections and water supplies. These are most likely to be found on existing rural industrial developments where the erection of a two storey structure with a flue stack will not be out of place.

Straw can be stored in a variety of locations, from existing on-farm facilities, to open stores on available land and to existing or purpose built covered storage facilities. The widespread use of straw as a fuel will naturally demand more storage facilities and this may cause some visual intrusion if not properly controlled.

Wet Agricultural Wastes

Animal slurries represent a major potential source of pollution, especially of water courses. In addition, where slurry handling facilities are present, odour can be a major problem, especially where a simple holding lagoon is employed without any additional treatment. Aeration of the slurry can reduce this problem. If left to decompose naturally, the slurry would yield carbon dioxide and methane (from the anaerobic portion of the slurry beneath the surface of the lagoon). Whilst the carbon dioxide emitted is only being recycled to the atmosphere, the methane is a potent 'greenhouse gas' and can be a burden on the environment. Anaerobic digestion remains the best method of avoiding these problems with the concomitant advantage of generating a usable energy fraction.

Environmental burden (units per kWh$_e$)	Straw	Forestry Wastes	Animal Slurries
Gaseous emissions:			
Carbon dioxide	1.7 kg	1.9 kg	1.1 kg
Sulphur dioxide	2.8 g[a]	1.9 g[a]	1.1 g[d]
Nitrogen oxides	2.7 g[b]	1.0 g[b]	2.0 g[e]
Carbon monoxide	4.2 g[c]	5.1 g[c]	2.6 g[c]
Volatile organic compounds	0.2 g[c]	0.2 g[c]	2.0 g[c]
Particulates	0.8 g[c]	1.0 g[c]	0.06 g[b]
Heavy metals	Trace	None	None
Ionising radiation	None	None	None
Solid waste	68 g[g]	15 g[g]	Approx. 1 kg[g]
Liquid waste	None	None	Approx. 10 kg[g]
Thermal emissions	4 kWh[f]	4 kWh[f]	2.3 kWh[f]
Amenity & Comfort:			
Noise	Minimal	Minimal	Minimal
Visual intrusion	Impact may be significant	Impact may be significant	Minimal
Use of wilderness area	None	Some impact possible	None
Other	Some transport	Some transport	None

Table 5: Environmental burdens. Atmospheric emissions on utilisation – based on combustion plant (wood, straw) & AD for slurry.
Notes   a: Derived from sulphur content
          b: Measured data
          c: New plant standards
          d: Inlet gas scrubbed
          e: German emission standards
          f: Thermal emission unlikely to be significant due to plant scale (some could be recovered as useful heat)
          g: May be used as a soil conditioner/fertiliser.

There are no environmental emissions requirements for digestion plant, but there are requirements on the handling of the potentially explosive methane produced by the process. These plant are silent in operation and present no visual disturbance as they are either completely or in part buried below the ground. However, if a gas holding facility is installed, it may be visible from a distance. A gas engine will be required for converting the gas to electricity; this can be housed in a low, sound proofed building. All these structures are of course likely to be located on the farmstead, and are therefore unlikely to look out of place.

Emissions

It is difficult to give a single set of emissions figures for "typical" plants. The information given in the following table is drawn from monitoring trials from a limited range of plant. No information is available for green wastes. New plant built under successive tranches of the NFFO is expected to become more efficient and better controlled. Gasification, in particular, will cause a step change in conversion efficiency and a dramatic reduction in emissions per unit of electricity generated.

In simple terms, all the carbon dioxide released when the agricultural and forestry wastes are combusted represents only that recently removed from the atmosphere during the growth of the biomass. The carbon is therefore being recycled. The emission burden associated with the production of the biofuel has not been included in the following calculations because the energy ratios are reasonably high and the burden associated with the equivalent production of fossil fuels will compensate to some extent in this comparative exercise. The plant thus displace all of the carbon dioxide emitted by the UK plant mix (1990) and displace a proportion of the sulphur dioxide and nitrogen oxides emissions. The figures in table 6 are calculated using the estimated mix of agricultural wastes plant in 2005 and the emissions in table 5. Further gains will be made by the adoption of gasification and pyrolysis.

Emission	Emission savings (grammes of oxide per kWh electrical)	Annual emission savings per typical 5 MW$_e$ combustion scheme generating 37 GWh/year (tonnes of oxide)
Carbon dioxide	734	27,000
Sulphur dioxide	8	300
Nitrogen oxides	1.3	50

Table 6: Emissions savings relative to average UK plant mix in 1990.

Health

There are no major safety risks associated with the harvest, transport and use of the biofuels described in this module. The only area of concern is associated with the storage of the fuels. Here, straw may represent a significant fire risk and appropriate recognition of this must be made when siting and planning the store. Storage of wood chips can lead to spore release which, in a confined space, represents a health hazard. Prolonged exposure to these spores can lead to the condition known as Farmer's Lung. This is easily controlled by wearing face masks as is currently the practice in Scandinavia.

E. Economic Performance

Combustion plant for wood fuel is well developed, but the conversion efficiencies at scales around 10 MW$_e$ will always be relatively low: approximately 20% to 25%. Gasification is likely to offer higher conversion efficiencies at similar capital costs and so improve the economic viability of wood fuelled power stations. Combustion technology will probably remain the preferred option for heat only applications. More information on wood gasification is given in the Advanced Conversion Module.

Straw combustion is problematic in comparison with wood, but equipment is available and has been successfully demonstrated in commercial projects in Denmark. As for wood combustion, conversion efficiency, even for the larger plants, is unlikely to exceed 25% without incurring relatively high capital costs of adding superheaters which may have to be heated separately with wood fuel. For these reasons, the cost of electricity produced by straw combustion plants is unlikely to fall as low as that generated by wood combustion for plants of the same capacity. Flash pyrolysis is being explored as an option for the future. If the pilot work is successful, then relatively small plants of less than 10 MW_e could be viable with conversion efficiencies of approximately 35%.

Animal slurries have a low dry solids content and are best suited to anaerobic digestion (AD). AD systems have been available for many years. The main problem is the high capital costs: running costs are extremely low. On-farm systems generating power for export are unlikely to be attractive. Centralised systems operating at a larger scale will probably prove to be the way ahead. Additional revenue streams may be tapped by adding a proportion of meat processing wastes to the slurry and by selling the fibre from the digester for soil conditioning. In addition, the sale of waste heat will probably prove necessary to make projects viable. More study is required to assess the markets for wastes, fibres and heat.

It is difficult to generalise about the economic performance of plants given the wide range of feedstocks and scales of operation. However, the main parameters for some typical plants available now are given below. They exclude insurance, rates and land rent.

Straw:
- Plant Description: Steam Cycle 12 MW_e

- Capital Cost: £1,500/kW_e

- Annual Running Costs: £260/kW_e

- Plant Life: 20 to 25 years.

Forestry Residues
- Plant Description: Steam Cycle, 6 MW_e

- Capital Cost: £1,400/kW_e

- Annual Running Costs: £330/kW_e

- Lifetime: 20 to 25 years.

Animal Slurries & Green Wastes
- Plant Description: Anaerobic Digester plus CHP, 100 kW_e (via spark ignition engine)

- Capital Cost: £4,500/kW_e

- Annual Running Costs: £80/kW_e

- Lifetime: 20 years.

F. Resource-Cost Curves   Figures 6 and 7 are for the Maximum Practicable Resource which takes account of the constraints upon the Accessible Resource and how they might change with time. These curves are consistent with the input data used in the energy systems modelling, the results of which are presented in Section H.

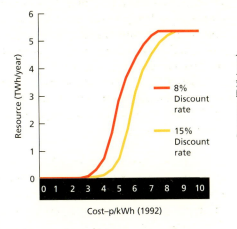

Figure 6: Maximum Practicable Resource-cost curves for agricultural wastes in 2005.

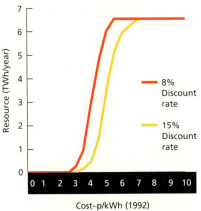

Figure 7: Maximum Practicable Resource-cost curves for agricultural wastes in 2025.

Forestry waste

Taking into consideration the location of the resource and the terrain from which it must be harvested, a conservative estimate of the Maximum Practicable Resource size will be around 35% of the Accessible Resource.

Straw

The exact size of the straw Maximum Practicable Resource is hard to estimate given the uncertainties over the long term acceptability of ploughing in as a viable disposal option, the optimum level of Set-Aside as a means of reducing cereal production and the availability of other markets, such as the livestock bedding and root vegetable top dressing markets which will in turn be influenced by the CAP reform. As a conservative estimate, the Maximum Practicable Resource for straw will be 30% of the Accessible Resource.

Wet Agricultural Wastes

The total Accessible Resource available from farm slurries is sufficient to generate 2.86 TWh/year. However it is clear that all of this resource is not available for digestion, and that the cost of the facilities necessary for slurry collection, handling and digestion will make exploitation of the resource uneconomic in the majority of single farms. However, as evidenced abroad , this situation can be reversed if larger, centralised plant are considered. Given the adoption of such plant and environmental legislation which is increasing the need for the farming industry to employ high quality slurry management systems, the Maximum Practicable Resource is likely to be as high as 50% of the Accessible Resource.

Maximum Practicable Resource		
Year	2005	2025
Straw	2.2	3.1
Forestry wastes	1.4	1.8
Animal slurries	1.4	1.4
Green farm wastes	0.3	0.3

Table 7: Maximum Practicable Resource potentials (TWh/year) at less than 10p/kWh for both 8% and 15% discount rates.

G. Constraints and Opportunities

Technological / System

With biofuels which are produced in remote rural locations, plant size will always be limited by the availability of the resource within economic transport distance. There is, therefore, little opportunity for reaping the benefits of economies of scale. Therefore the deployment of these biofuels is driven by the lower capital costs which may be derived from a move towards high efficiency advanced conversion technologies. For small plant, there may be scope for improving the economics if the plant is used to provide heat for a combined heat and power scheme. If biofuels could provide an economic heat source in rural areas, this would obviously prove an asset to the rural economy.

An additional burden imposed by siting generation plant in rural locations is the availability and cost of suitable grid connection. However, a possible benefit here is that in certain circumstances embedded generation may obviate the need for grid strengthening. In addition, generation from these biofuels will be at a constant base-load, which will complement the supply from more intermittent sources of renewable energy such as wind power.

With both straw and wood fuel, the trend is for the industry which supplies the fuel (farming and forest products respectively) to adopt increased levels of mechanisation, this will ultimately reduce the delivered costs of the fuels. In the case of straw fuel, this move has resulted in the adoption of high density big bales which reduce transport and storage costs. This development is limited because the current markets for straw ultimately require loose straw. It will only be the energy market which requires high density straw, and it is only when the first markets for the straw are created that the industrial development of suitable systems will occur. In this way, adoption of systems similar to those used to compact refuse on dustcarts may find application.

The forest harvesting industry, whilst adopting increasing levels of mechanisation, will not adopt systems leading to reduced fuel wood harvesting costs (integrated harvesting methods) until the fuel wood market is available to justify the high levels of investment required. However, all the machinery is available commercially and little further development will be required in the short term. In the future, commercial pressures will bring about improvements and developments as necessary.

Both wood and straw combustion plant are based on proven technology and present little commercial risk. The only problem is that the low ash fusion temperatures for straw fuel may lead to slagging problems and increased maintenance costs. If a market is created for these plant in the UK, more UK manufacturers will be encouraged to produce them. This increased competition will lead to further downward pressure on the capital cost of conversion plant. In the future, the obvious advantages of increased conversion efficiency, linked to the potential for lower capital costs, will hasten the adoption of advanced conversion technologies such as gasification. These systems are now entering the commercial demonstration phase.

Institutional

As with any technology which is perceived as being novel, project financing can be a problem. In addition, the utility developing the project is often looking for rates of return in the order of 25% as a result of the perceived risk associated with novel technologies. Unfortunately technologies which have been proven overseas but have not yet been demonstrated in the UK are also perceived as risky. This is true of all of the dry waste processes described in this module, as well as of large scale anaerobic digestion of animal slurries and green crop wastes.

New employment opportunities accruing from the adoption of this technology will almost all be in the rural sector. This is because all the fuel is produced in rural areas and must be used close to the site of production to minimise transport costs.

Access to the resource does not appear to present any problems, as these biofuels are all derived from existing forestry cropping or farming activities. In addition, management of these wastes will bring with it significant environmental advantages which are likely to be welcomed.

Legislation governing combustion processes is a matter of concern across all the biofuels programmes. If these become too stringent, the cost of generation may become prohibitive – especially compared with generation from fossil fuel sources which do not carry such a great legislative burden. For animal slurries, and to a lesser extent green crop wastes, environmental legislation is likely to be the driving force behind the adoption of high quality slurry handling mechanisms, including anaerobic digestion.

The expertise necessary to produce the biofuels discussed in this module is available in the UK to a greater or lesser extent. The area of greatest concern in this regard is the provision of fuel wood from conventional forestry. Here, the new mechanisation required is not in general use and new skills must be learnt. This may also increase the perceived risk of projects. Nevertheless, the forest harvesting industry is moving towards more mechanisation and this problem may therefore prove a minor one.

In terms of fuel utilisation, it is clear that whilst most combustion plant is similar in design to existing sloping grate coal combustion technology, the expertise necessary to operate this kind of plant is not widely available at present. In addition, the area of fuel handling and storage is new to British industry, though these skills will be quickly learnt. Experience to date in operating on-farm anaerobic digestion plant has not highlighted any areas where expertise is lacking, or cannot be easily acquired.

H. Prospects Modelling Results

Scenario	Discount rate	Contribution: TWh/year			
		2000	2005	2010	2025
HOP	8%	0.14	0.22	0.13	0
	15%	0.14	0.22	0.22	0
	Survey (10%)	0.14	0.22	0.13	0
CSS	8%	0.22	0.22	0.22	0
	15%	0.22	0.034	0.22	0
	Survey (10%)	0.22	0.22	0.22	0
LOP	8%	0	0.034	0.034	0
	15%	0	0.050	0.034	0
	Survey (10%)	0	0.22	0.034	0
HECa	8%	2.7	5.2	5.2	6.5
	15%	2.7	5.2	5.2	6.5
	Survey (10%)	2.7	5.2	5.2	6.5
HECb	8%	2.7	5.2	5.3	6.5
	15%	2.7	5.2	5.2	6.5
	Survey (10%)	2.7	5.2	5.2	6.5

Table 8: Potential contribution under all scenarios.

From table 8 it can be seen that under the Composite (CSS) scenario, the contribution from these energy sources is limited to those projects supported under the NFFO. The contribution is much larger in both of the Heightened Environmental Concern (HEC) scenarios. These are close or equal to the Maximum Practicable Resource potentials shown in figures 6 and 7. These contributions appear independent of discount rate, indicating that plant cost and efficiency constraints are not limiting the uptake of the technology. The contribution increases after the year 2000 with the availability of higher efficiency gasification plant.

Both the high and low oil price scenarios (HOP & LOP) fail to elicit a contribution from agricultural wastes as high as that obtained under even the CSS scenario. This indicates that, even though demonstration generation projects are available, MARKAL has chosen not to utilise this plant fully due to the relatively high variable costs involved. The HOP scenario shows an interesting trend with more of the available capacity being used from 2010 under a 15% discount rate regime than under an 8% discount rate regime.

Environmental burden			Incremental contribution: TWh/year			
	Constraint	Base case scenario	2000	2005	2010	2025
Carbon dioxide	- 5%	CSS	0	0	0	0
	- 5%	HOP	-0.10	-0.12	0.022	0
Carbon dioxide	-10%	CSS	0	0	0	4.1
	-10%	HOP	0.089	0	0.089	4.1
Sulphur dioxide	-40%	CSS	0	0	0	0
	-40%	HOP	0.089	0	0.089	0
Nitrogen oxides	-40%	CSS	0	0	-0.22	0
	-40%	HOP	-0.01	-0.089	-0.13	0

Table 9: Incremental contribution under environmental constraints.
(TWh/year above base case scenario contribution).

Table 9 shows that the major influence on the adoption of agricultural wastes as a fuel source is the need to reduce carbon dioxide emissions by 10% at 2025. This is achieved by utilising around 75% of the Maximum Practicable Resource available. Interestingly, achieving a 5% saving does not require the adoption of agricultural and forestry wastes as fuels.

The potential for biomass fuels to give higher emissions of nitrogen oxides results in a modest reduction in contribution under the 40% reduction in emissions of nitrogen oxides under HOP scenario. However, as the preferred utilisation option by this date will be gasification, there will be the opportunity to reduce emissions of nitrogen oxides from the levels reported in table 5 (and used in the MARKAL model), which are based on traditional combustion plant.

Discount rate	Incremental contribution: TWh/year			
Year	2000	2005	2010	2025
8%	0	0	0	0
15%	0	0.19	0	0.9
Survey	0	0	0	4.1

Table 10: Incremental contribution under Shifting Sands assumptions.
(TWh/year above CSS contribution.)

Under the Shifting Sands scenario the only significant additional contribution occurs at the survey discount rate (10%). The reason why this is the only discount rate producing such a contribution, or why at a lower 8% discount rate the effect is zero, is the availability of capital intensive technologies with low variable costs at the lower discount rates.

II. Programme Review

A. United Kingdom

The technologies described in this module are at the point of commercial exploitation in other countries. The purpose of the Government's programme is to create a similar situation in the UK.

A major barrier to be overcome in this regard is the lack of a market for renewable fuels derived from agricultural and forestry wastes. This brings with it a lack of investment in appropriate mechanisation to lower the delivered price of these fuels. Once the fuels derived from agricultural and forestry wastes are recognised by the relevant supply industries as another crop to be sold into a competitive market, the cost of utilisation plant will not be seen as a barrier. Commercial exploitation of these fuels will occur where they are competitive with fossil fuels.

The Government's Programme
Since the last energy paper there has been considerable research into methods of deriving and using fuels from agricultural and forestry wastes.

Forestry Wastes
Harvesting trials have been completed using the latest range of mechanisation options imported from around the world, both in isolation and in combination, to produce a wood fuel chip as a second crop when harvesting a range of tree species grown on a range of forest terrains. From this work a computer decision support system has been developed to predict the delivered price of fuel from the crop input data. Central to this programme has been international collaboration through participation in the IEA Bioenergy Agreement.

As a direct result of the DTI's Wood as a Fuel Programme, one pass, integrated harvesting operations have been devised for UK conditions. These have yet to be commercially demonstrated.

During 1993, the work on the supply of wood chip fuel was expanded into a number of studies investigating the techno-economic feasibility of constructing wood fired power stations under the NFFO. These studies have been carried out with industry and are awaiting the third NFFO order before the decision to proceed is made. The harvesting trials undertaken to date will need to be developed into demonstrations of integrated harvesting practice.

Straw
Since the last energy paper review, considerable efforts have been made to investigate means of densifying straw effectively at low cost. This will give lower transport and storage costs and allow existing coal technology to be used for fuel feeding and combustion. This work has shown that densification, whilst technically feasible, requires large plant (precluding in-field densification as an option) and a significant energy input. This is a major drawback.

Developments overseas have led to the introduction of 'cigar burner' technology in which normal bales are fed into the combuster and the straw fuel combusted from the end. This has been adapted in the UK by one combuster manufacturer. This activity has been augmented by involvement in the IEA Bioenergy Agreement.

During the past year, the work on the supply of straw fuel has been expanded into a number of studies investigating the techno-economic feasibility of constructing straw fired power stations under the NFFO. These have been carried out with industry and are awaiting the third NFFO order before the decision to proceed is made. At this point, the trials undertaken to date to create suitable supply strategies will need to be further developed.

Farm Wastes
Under the NFFO a small number of commercial demonstrations of the utilisation of anaerobic digestion of farm slurries to generate biogas for electricity generation have been undertaken. These, allied to technology status reviews, have highlighted the poor economic performance of this option for energy production alone. However, international collaboration via the IEA Bioenergy Agreement has made it clear that centralised slurry handling facilities have a better economic performance allied to lower potential environmental impacts. Future DTI involvement in this area will be aimed at replicating such developments in the UK.

B. Other Countries

Forestry Wastes
The use of wood fuel is commonplace in many countries. The scale of operation varies enormously from domestic heating through to large scale power generation. In northern Europe, domestic scale heating is widespread, especially in more remote areas. In towns, district heating schemes fuelled by wood are common. The former activity reflects the traditional practice of collecting fuel wood from forests which are widespread in these regions. The operation of district heating plant on wood often reflects the fiscal incentives in place in countries such as Denmark and Sweden which make fossil fuels for domestic consumption expensive through taxation. Power generation either through dedicated plant or CHP schemes is not common in these countries. However, the combustion of wastes from sawmilling, paper making and particle board manufacture at the plant to provide energy for the manufacturing process is carried out almost universally.

In North America whilst domestic fuel wood consumption is again relatively common, especially in rural areas, there are also large scale power generation projects in place totalling over 6,000 MW$_e$ in the USA alone.

Research into wood fuel production is still being undertaken in the Scandinavian countries, but this mainly in response to environmental concerns. As a result, 'low impact' harvesting techniques are being developed. In North America, whilst the energy market is relatively large, research into fuel wood production is not a priority. This in part reflects the fact that a proportion of the residues from large scale logging operations are available cheaply negating the need for specialist fuel wood harvesting methods. Only if the fuel wood market is to be expanded will pressure to research supply-side issues develop.

Utilisation research reflects the extremes of markets in which fuel wood competes. On the small scale, there is concern regarding the pollution from log burning domestic stoves. This is of particular concerns in countries like Switzerland where high levels of smoke production in mountainous regions is causing smog problems. Thus much work is being undertaken looking at the efficient combustion of wood fuel in stoves. Where countries are utilising wood fuel on the large scale for power generation, research into gasification processes is being undertaken. In addition, in the USA there continues to be work looking at the production of methanol from wood as a transport fuel.

Straw

Denmark is the only country in which large scale utilisation of straw for energy is undertaken, though this option is being considered in others such as the USA. In Denmark, the energy produced is mainly used in district heating plants, although power generation via CHP schemes is also in commercial operation. In some of these systems the straw is co-fired with wood or coal.

The reason why straw as an energy source is being exploited in Denmark is due to the tax on fossil fuels. Originally this taxation was introduced as a result of the energy crisis in the 1970s driving up the price of fossil fuels. At that time – prior to the development of the Danish gas fields – almost all Denmark's fossil fuels were imported. The tax is now retained for environmental reasons.

Much of the R&D being undertaken in Denmark is directed towards the utilisation of straw, rather than on production technology. This is because the fuel is harvested and handled using existing farming equipment. However, some work is being undertaken to reduce the delivered price of the straw.

Farm Wastes

Anaerobic digestion of animal slurries to produce energy on a large scale is being undertaken in Denmark, and to a lesser extent Holland. Interest in this technology is also being stimulated in the USA and Switzerland. This is in addition to facilities to handle MSW developed in other European countries.

In Denmark there are currently nine facilities handling farm wastes, with one utilising mainly MSW. These sites also accept other organic wastes to a greater or lesser extent. These wastes attract a disposal fee which improves the economics of the system.

Research in this area is directed at employing thermophilic process to improve the gas yield and reduce the residence time in the reactor.

Green Agricultural Wastes

Crop residues are often utilised for fuel in countries producing crops with a dry woody residue such as coconut husks, or the bagasse from sugar cane. There is current interest in utilising these wastes as fuels in gasification plant for power generation, or CHP duties.

Green, non-woody residues are not utilised routinely as fuels. In developed countries, or those with indigenous energy supplies, the cost of utilising these materials as fuels does not make this an attractive option. In less developed countries, the value of green wastes as animal feeds or green manure outweighs the potential benefit of using this material as a fuel.

As in the UK, there is potential to co-digest this biomass with animal slurries, and this is an increasingly likely option in those countries employing anaerobic digestion technology.

Energy Crops

I. Technology Review

A. Technology Status

Fuels may be derived from crops in solid, liquid or gaseous form. Historically, there has been limited direct development of energy crops, with the exception of coppice to provide fuel for charcoal production – but this was not grown as an agricultural crop. Similarly, the means to harvest or convert energy crops into biofuels has not been developed either. However, some annual food crops have been used with standard farm methods of cultivation and harvesting.

The methods of converting crops to biofuels are varied and reflect the diversity of the biomass resource. The drier, higher lignin content materials are best converted by thermal processing (combustion, gasification or pyrolysis) whilst the wetter biomass can be converted to either a liquid or a gaseous fuel through biological processing (fermentation under anaerobic conditions). In the latter case, anaerobic digestion will yield a methane-rich biogas of medium energy content. Other fermentation techniques can yield liquid fuels, the most important of which is ethanol. More recently the conversion of rapeseed oil via esterification to a diesel substitute has received much attention.

For both dry and wet materials, processing and use may be complicated by the incorporation of other material such as wastes (see Agricultural Wastes module). For the drier materials (predominantly wood) a typical project might involve the growing of a closely spaced (10,000 trees/hectare) arable coppice crop of willow or poplar by a co-operative of several farmers. (One hectare (ha) = 10,000 sq.m). The crop on each farm will be grown at whole field scale (approximately 10 ha), but the extent of the area used will depend on the size of the market being serviced. Good agricultural land is preferable since this will give the greatest yield.

Figure 1: Coppice plantation.

Willow is best suited to a two to three year harvesting cycle and poplar to a two to five year cycle. The latter, on a five year rotation, requires less dense planting (approximately 6,700 trees/ha) and will be adopted where existing forestry equipment and expertise is available. In each case the average yield is

Figure 2: Miscanthus.

in the range 10 to 15 dry tonnes/ha/year. The harvested material would then be collected, transported and stored close to the point or points of use. During storage (which can be of cut sticks or chipped material) some drying will occur either naturally, or by way of forced draught ventilation. This will reduce the microbial breakdown of the wood and also increase its calorific value.

The C4 grasses (which photosynthesise more efficiently than native C3 plants) such as Miscanthus, are very similar to coppice in that they are perennial, harvested in the winter using similar machinery, and the chipped product produced requires the same infrastructure for storage and transport. The fuel has a similar calorific value per unit weight as wood and can be utilised in the same plant. The potential advantages offered by these crops are that at harvest the fuel is relatively dry (most of the stem moisture being translocated into the rhizome) and that the yield potential is significantly higher (15 to 30 dry tonnes/ha).

Combustible biomass fuels (e.g. wood and Miscanthus) are relatively low quality fuels having a calorific value around 19 GJ/dry tonne. The calorific value of these fuels falls linearly with increasing moisture content such that at 55% moisture by weight (the typical moisture content at which wood is harvested) the calorific valve is around 10 GJ/tonne. They have very low sulphur contents and give only 1% to 2% ash. A range of combustion plant is available specifically designed for wood fuel; this should also be suitable for Miscanthus. Available plant starts at the wood-burning stove level, but automatic systems to combust chips start at around 0.1 MW_{th} and can be as large as the market requires. At the 0.5 MW_{th} size upwards, this equipment can efficiently combust fuels with a moisture content of 55%. Smaller plant typically require drier fuel. Plants of this kind are well proven.

A typical sized plant is hard to define since fuel availability will be a determining factor. Due to the high cost of transporting these fuels, the catchment area for a plant is unlikely to exceed a radius of 50 km. As a result, a typical installation fuelled by energy crops will generally generate less than 50 MW_e. The availability of additional resource from conventional forestry could increase the size of the project and, in some circumstances, dual fuelling with straw may be possible. Opportunities of this kind will vary from region to region but are likely to arise predominantly in East Anglia where both straw and arable land available for energy crops are in abundance.

Another attractive option is co-firing with fossil fuels. By introducing a percentage of biomass into a conventional power station high conversion efficiencies are achieved and many of the costs (infrastructure, grid connection, capital costs of buildings and labour) are shared. So the energy crops input could be a few percent of a station generating a few gigawatts. However, with this approach, the wood ash is no longer useful as a fertiliser: by co-firing with fossil fuels the ash becomes a disposal problem.

Higher efficiency utilisation of these resources is now becoming possible through the use of gasification plant. This technology is currently entering the commercial demonstration stage.

The production of liquid fuels from crops can be achieved through two routes. High carbohydrate crops (cereals or sugar beet) can be produced as feedstocks for fermentation processes yielding ethanol as a petrol substitute. Rape seed can also be used as a feedstock and in this case the oil produced is esterified into a diesel substitute. These are both simple processes. After production, the alcohol fuels may be blended with petrol as extenders or octane enhancers rather than used as direct replacements for petrol. Direct substitution will require extensive engine modifications. Biodiesel on the other hand can be used in existing unmodified engines with little, if any, adverse effects. In both cases, crop production would be the same as for food production and existing transport and distribution networks would be utilised for both the crop and the liquid biofuel product.

Plant producing the esterified rape oil in operation overseas are in the size range 15,000 to 60,000 tonnes of fuel output per year. These plant require feedstock from 12,500 to 50,000 ha respectively.

Digestion of green crops to produce energy will be as for green agricultural wastes; this is described in the Agricultural Wastes module.

Combustion plant have an operating life of some 25 years, after which time they can be replaced or decommissioned with the scrap value covering most of the costs. There is also a need to consider the effects of reversion of land use back to food crops. Where crops are annuals any change away from fuels could be accommodated relatively easily. For arable coppice and the perennial C4 grasses, the situation is complicated in that the crop has a life expectancy of 30 years and premature change of use will require tree stump or root removal before other uses can be made of the land. However, this is a relatively easy operation (unlike conventional forestry) and the land can return to food cropping in one season. Storage silos and buildings can be converted easily to other agricultural use. Indeed, these buildings are likely to have been converted from agricultural use in the first place to accept the products of energy cropping.

B. Market Status

No commercial energy-crop-to-fuel projects currently exist in the UK. There are however, significant developments overseas for example, the bioethanol programme in Brazil and the biodiesel plants in Austria, which produce these biofuels commercially. Also, in Sweden arable coppice is in large scale production to fuel district heating plants. Currently in excess of 6,000 ha have been planted.

In addition to these commercial developments, there is considerable interest in developing novel crops and appropriate conversion technologies both at home and abroad. In the UK arable coppice is being developed with Department of Trade and Industry funding. Extensive programmes are also under way on various crops in Scandinavia, Europe (particularly through DGXII and DGXVII of the CEC) and North America. In addition, the International Energy Agency (IEA) has a substantial collaborative programme.

The crops with the greatest long term potential are those which produce a combustible product, the principal ones being arable coppice and the C4 grasses. They produce a better energy balance than annual energy crops and thus stand a better chance of becoming economically viable without government support. There is a problem, however, in creating a market for these crops, unlike those which produce liquid biofuels which have a ready market if their economics look favourable. The current programme, therefore, is focused on developing the market and growing these combustible crops. Utilisation of crops may well begin in combination with agricultural and forestry wastes. In successive NFFO tranches, the capacity of biomass plants is likely to increase together with the proportion of energy crops used as fuel. Economies of scale coupled with decreasing energy crop production costs, improving conversion techniques, and the development of a wood fuel market will allow the NFFO support price to decrease with time. Medium to long term projections indicate that energy crop power stations will be able to compete with new-build coal plants without the NFFO.

Some deployment of this technology may be achieved on a small scale by accessing heat markets in the rural community, close to the site of crop production, where the price of competing fuels is highest. The advantages of this would be that in rural and farming communities people already have, or can easily develop, the necessary supply infrastructure. Once a market of this kind is established the development of more effective harvesting, transport and storage methods should occur without further inducements.

Abroad, considerable emphasis is being placed on the production of liquid biofuels despite their low energy ratios and poor economic performance compared with other energy crops. First, utilising food crops for non-food use, especially where the market into which they are being sold is large, represents a 'quick fix' for the problems of declining farm incomes and food over-production. Secondly, liquid biofuels are biodegradable. This is particularly important in environmentally sensitive areas such as inland waterways and on scenic mountain slopes used for skiing Lastly, the strong moves to decrease emissions of sulphur in city centres has increased interest in biofuels. The last two reasons have obvious benefits which may justify production and use of liquid biofuels.

	Combustible Crops	Crops giving Liquid Biofuels
UK installed capacity	None	None
World installed capacity	None utilising crops alone, but in conjunction with forestry wastes, many MW$_{th}$ in Sweden.	Biodiesel from oilseed rape produced in Austria, Italy and France. Bioethanol produced in Brazil, USA and France.
UK industry	Ideally placed to both produce and utilise the crops.	Ideally placed to both produce and utilise the crops.

Table 1: Status of technology.

C. Resource

For the majority of energy crops the major constraint on the potential is that of land availability. The Department of Land Economy at Cambridge University estimates that 1.0 to 1.5 million ha of land may become surplus to requirements for food production by the year 2000. By 2010 this figure is expected to rise to 5.5 million ha. All of this land is potentially available for the growing of energy crops. (Currently 4 million ha of land are under cereals out of a total of 18.5 million ha used for agriculture.)

The European Commission introduced a CAP Reform programme during the autumn of 1992 because of the mounting food surplus and the consequent financial burden. On the arable side, to compensate for reduced support prices, farmers receive an area payment on eligible arable crops. Part of the reform package also entails setting aside some productive arable land. The amount of set aside was 15% of eligible arable area which was rotated around the farm to avoid only poorer land falling into the scheme. As of the 1994 harvest, it is anticipated that 18% of eligible area would need to be set aside to qualify for the non-rotational 5 year scheme (this figure is likely to be revised upwards after the 1996 harvest). Clearly, this scheme is applicable to perennial crops such as short rotation coppice or wood grasses such as Miscanthus. On the livestock side, existing subsidy regimes will be subject to a mix of maximum stocking rates and eligible headages. This may lead to surplus land, although in practice it is likely to lead to more extensive grazing systems. Overall, existing and potential changes to the CAP are likely to release more agricultural land from food production which could then be used for energy crops.

ource	Yield (GJ/ha/year)
oppice wood	190 to 399
Miscanthus	270 to 720
Digestible crops	135
Whole crop cereals	126 to 145
Rape	Oil only – 49 Whole crop – 110

Table 2: Variation in yield expected for each biofuel derived from energy crops.

Yields of other energy crops will vary enormously, the best being fodder crops and whole crop use of cereals grown as two crops within one year and producing typically a total of 10 dry tonnes of biomass per ha/year. The wetter crops would then be converted to biogas or ethanol, the drier materials being burned, pyrolysed or gasified. Novel crops such as the woody grass Miscanthus have been shown to give high yields (approximately 20 dry tonnes/ha/year harvested annually) in field plots located in Germany, in areas with a similar climate to the UK.

Some energy crops such as coppice have additional value as game cover. As a result they may be produced for non-energy reasons and thus the harvested fuel may be available at low cost.

England & Wales	Scotland	Northern Ireland	UK
139	23	32	194

Table 3: Accessible Resource (TWh/year) in the UK at a cost of less than 10p/kWh. Note: this split is based on a simple proportioning of the available farmland in each of these regions. In practice it will be the CAP reforms and the way they relate to the various farming enterprises (in each region) that will dictate the exact breakdown geographically.

In estimating the Accessible Resource for arable energy crops the following assumptions have been made:

- all surplus land is used for these crops (1.0 M ha by 2000 and 5.5 M ha by 2010, giving a linear interpolation figure of 3.25 M ha for 2005 and a maximum limit of 5.0 M ha for 2025, which accounts for 0.5 M ha of conservation headlands and field margins);

- average yield of all arable crops will be 15 dry tonnes/ha/year by 2000 and 21 dry tonnes/ha/year by 2010 through adoption of improved clones produced from, say, a willow breeding programme;

- a conversion efficiency of 35% has been used which assumes the introduction of gasification technology.

The surplus land figure is difficult to define and calculate. It is subject to continual review in the light of rapid changes in agricultural policy and practice.

In interpreting table 3 it must be borne in mind that the land available for energy crops can only be used for one crop at a time; thus land given over to growing arable coppice is not available for growing Miscanthus.

Figure 3: Harvesting willow.

D. Environmental Aspects

When considering the constraints on realising this potential it must be remembered that energy crops will be produced on land which is currently being used for food production – 'new' land is not being utilised. Incentives will be required to bring land out of food production, to promote energy crops and to establish a stable market and maintain farm incomes.

Apart from the visual impact of growing the perennial crops such as coppice and the C4 grasses, there may also be an impact on the local water table with coppice, due to the intensity of tree planting. However, the willows used in energy forestry are typically shrub, not tree, willows and do not require wet conditions. On the other hand, crops such as Miscanthus may be more efficient in their use of water than existing crops; this may be an advantage in certain areas of England.

Figure 4: Mixed clonal coppice plantation.

Water aside, these crops do have a less demanding appetite than food crops for other nutrients from the soil – about a fifth of the annual requirement of cereal crops, for example. A further advantage is the possible use of the crop and its fibrous root structure to absorb nutrients from sewage and other sludges by acting as a biofilter. This is potentially very beneficial and could markedly improve the prospects for energy coppice or Miscanthus in nitrate-sensitive areas like East Anglia where it could prevent leachate run off and associated eutrophication.

Siting of energy crop production projects will occur in rural or semi-rural positions. The arable coppice and C4 grass crop will be extensive but not over-intrusive since it will be grown like other agricultural crops in traditional field settings. The height of arable coppice, particularly during its third and final growing season before cropping, will be several metres. It will be less intrusive in winter, however, since it is deciduous. Due to the long production cycle of coppice and the use of rotational planting, the general visual appearance will be stable but there will be continual variations within the crop.

For both coppice and C4 grass crops harvesting will be carried out in winter. Most other management tasks will also be undertaken then, the exception being the application of sludges or other nutrients which must be carried out with the trees in leaf. Here, additional vehicle movements will be necessary but it is believed that generally these will still be less than those required for conventional cropping.

If no existing grain storage facilities are available near to the point of use of the derived fuel, silos for the storage of wood chips may need to be installed. Buildings will be needed to house the boiler and generator equipment and, if electricity is to be exported, the necessary interconnections will have to be added.

Crops destined for biological processing will be grown as arable crops – many of these are standard farm crops such as wheat and sugar beet. There will be no obvious changes, therefore, at the production or harvesting stages. However, for processing, substantial silos will be required to contain the fermenting materials and these will be difficult to disguise. They will be similar to normal agricultural buildings but, if necessary, their impact could be reduced by partial burial.

The production and processing of rape oil for fuel will be much the same as for food. The only additional requirement will be the construction of biodiesel factories to process the oil into the rape methyl ester. From this point a separate additional distribution network for the fuel may be required to run in parallel with that for fossil diesel.

Planted as a monoculture, any crop has the potential to develop and harbour pests and diseases. Of particular worry at the present time is the fungal rust disease which is pandemic amongst most willow clones in the UK. As with other crops, mixed clone or even species planting may well be required to minimise such problems. Also there will be the need to introduce new clones continually. The introduction of natural 'biological' control is a likely medium term solution.

A positive impact of arable coppice will be the greater diversity of flora and fauna produced both within and at the margins of the crop. This diversity will be partly brought about by less frequent disturbance of the crop and land, since cropping will occur only once in every 3 to 5 years. Also since cropping will be in rotation, the whole field will not be cleared as in the case with food crops; this will give a greater opportunity for wildlife to take advantage of the substantial ground cover. The situation with the C4 grasses has yet to be determined but, since they are non-native species, these crops may not afford such good opportunities for biodiversity as coppice.

The margins around the perennial crops and crop corridors (broad open areas which can be allowed to develop natural vegetation) are essential to allow access for mechanical harvesting and are unique to this kind of crop (unlike 'wall to wall' cereal cropping as currently practised). These will provide useful habitats and the close proximity of cover afforded by both the crop and hedgerow will boost the local ecology. Indeed, it is possible that up to 10% of the field area will be taken up with such areas. Thus nationally by 2010 up to 550,000 ha of new habitat could be formed.

Generally the plant for converting energy crops to biofuels can easily be blended into a rural or semirural location. Perhaps the most intrusive aspect of combustion facilities will be the stack or chimney, the size of which will be related to site location and conditions (the requirement in this respect will be similar to that for a fossil-fuelled plant). For wet processing the fermentation silos will be the most obtrusive structures, but these will be similar to existing farm silos. Likewise, silos for wood chip storage may be large but will be virtually identical to grain and other silos used on farms.

Gaseous emissions from combustion facilities will be low in pollutants such as nitrogen oxides and sulphur dioxide (see table 4). Solid residues (the ash) will be rich in phosphates and potash and can be returned to the land to recycle these nutrients.

Emissions data for a fairly low efficiency forestry waste combustion plant are detailed in table 4. The emissions are expected to be lower for larger modern combustion plant and to be very much lower for gasification plant (see the Advanced Conversion module). The emissions per unit of electricity will fall in proportion with the conversion efficiency increases projected for gasification. In addition, this advanced conversion technique is inherently cleaner, giving further emissions reductions.

Environmental burden	Specification (units per kWh$_e$)
Gaseous emissions:	
Carbon dioxide	1.9 kg
Sulphur dioxide	1.9 g a
Nitrogen oxides	1.0 g b
Carbon monoxide	5.1 g c
Volatile organic compounds	0.2 g c
Particulates	1.0 g c
Heavy metals	None
Ionising radiation	None
Wastes:	
Solid waste	15 g solid waste (may be used as a fertiliser)
Liquid waste	None
Thermal emissions	4 kWh d
Amenity & Comfort:	
Noise	Minimal
Visual intrusion	Impact may be significant
Use of wilderness areas	Little impact
Other	Some due to transport

Table 4: Environmental burdens. Atmospheric emissions on utilisation, based on combustion plant. (Gasification plant will give much lower emissions per unit of electricity).
Notes a: Based on sulphur content
b: Measured data
c: New plant standards
d: Thermal emissions unlikely to be significant due to scale of plant (some could be recovered as useful heat).

Table 5 shows the emission savings relative to the average UK plant mix (1990). This assumes that, in simple terms, all the carbon dioxide released when the energy crops are combusted represents only that recently removed from the atmosphere during the growth of the biomass and that the carbon is being 'recycled'. Obviously, the balance is not perfect as fossil fuels will be burned in harvesting, transport, hardware fabrication etc. However, the energy ratio for short rotation coppice, taking account of all these inputs, is sufficiently high (greater than ten under most scenarios) that in this calculation the slight carbon dioxide imbalance can be discounted. In the same way, there is an emissions burden associated with the production of fossil fuels and this will tend to compensate for the energy crop production emissions in this comparative exercise.

Emission	Emission savings (grammes of oxide per kWh electrical)	Annual emission savings per typical 5 MW$_e$ combustion scheme generating 37 GWh/year (tonnes of oxide)
Carbon dioxide	734	27,000
Sulphur dioxide	8.1	300
Nitrogen oxides	2.4	90

Table 5: Emissions savings relative to average UK plant mix in 1990.

For sulphur dioxide and nitrogen oxides the emission saving is taken as the difference between that emitted by the average fuel mix and that emitted by wood firing as shown in table 4. These savings will increase with improvements to conversion efficiency and plant operation, especially with a move to gasification.

Anaerobic digestion processing facilities will produce a methane-rich biogas which is a relatively clean fuel producing low levels of nitrogen oxides and sulphur dioxide on combustion. Digestion also produces valuable residues which are both liquid and solid. These are rich in nutrients and can be used as soil conditioners.

The production and storage of quantities of methane-rich biogases (flammable and asphyxiating gas mixtures) requires precautions to be taken to ensure safe handling. Sparkfree apparatus and other safety measures are prerequisites. Likewise with the storage of wood chips or wood bundles in quantity, precautions against fire must be taken. Safe wood chip storage practices must be employed since on storage, wood chips begin microbial decomposition releasing large amounts of spores which, if inhaled, can lead to the condition known as Farmer's Lung. There are also some reports of rape pollen causing allergic reactions. This has yet to be fully quantified, but it may cause a problem if large scale production for energy is added to food crop production.

Figure 5: Wood chip storage

E. Economics

The range of processes and technologies is too varied to be covered here in detail. However, a broad summary of costs is given for the major options. This is supplemented by detailed cost and performance data for the most important option, the growing and use of arable coppice as a fuel for electricity generation. Both conventional combustion and gasification plant are considered.

In addition to this main option, fuel wood from arable coppice has the potential to become a valuable fuel resource in rural communities where oil, liquid petroleum gas and electricity are the major energy sources. Being small scale energy users the cost of their energy is high at approximately £2 to £4/GJ for oil, £6 to £7/GJ for LPG and £6/GJ for off-peak electricity. Coal prices are typically over £2/GJ in these areas. Arable coppice may compete as a combustion fuel for heat-only or for process steam-raising applications. The cost of wood combustion plant reflects not only the size of plant, but also its use. These prices are current, but reflect both the small market currently available for this equipment and the fact that it is all imported. In the future a larger market will introduce greater competition and downward pressures on prices. However, the following data can be used for indicative purposes.

Capital costs	£1,500/kW$_e$
Annual running costs & fuel	£300/kW$_e$/year
Plant availability	85%
Construction time	1.5 years
Lifetime	20 to 25 years

Table 6: 6 MW steam cycle plant at current prices and performances excluding insurance, rates, land rent.

Capital costs	£1,500/kW$_e$
Annual running costs & fuel	£200/kW$_e$/year
Plant availability	85%
Construction time	1 year
Lifetime	20 years

Table 7: 5 MW gasification plant at current prices and performances excluding insurance, rates, land rent.

A 0.3 MW$_{th}$ plant to produce hot water for space heating costs around £13,000. This compares with £100,000 for a 0.5 MW$_{th}$ system to give low pressure steam. Economies of scale mean that a 1 MW$_{th}$ plant for the same purposes cost only £30,000 and £150,000 respectively. Similarly, 5 MW$_{th}$ and 10 MW$_{th}$ low pressure steam heating plant cost £250,000 and £400,000 respectively. At the 10 MW$_{th}$ level, CHP becomes a viable proposition. Here, plant to raise 250 psi steam (575 kW electricity production) costs £840,000 while plant to give steam at 460 psi (1.13 MW electricity production) costs £1,700,000. It should be noted that whilst this plant is available now and is well proven, gasification is likely to be preferred in the future because of the higher conversion efficiencies that can be attained.

The gasification plant described in table 7 is based on technology which is available now and soon to be demonstrated commercially. In the future, higher conversion efficiencies and lower capital costs might be expected. The Advanced Conversion Technologies module gives more details.

To obtain the maximum energy from wet energy crops, they must be biologically processed. Anaerobic digestion produces methane-rich biogas which can be combusted in a boiler to raise hot water or steam. Alternatively the biogas can be used in modified spark ignition engines for electricity generation or combined heat and power. Anaerobic digestion is capital-intensive. A typical digester for a 1 MW$_e$ system with a spark ignition engine and generator will cost approximately £3.5 M to £4.0 M.

Wet or dry crops may be fermented to ethanol, a liquid fuel which can be used as a petrol additive or extender. Commercial scale plants based on wheat, maize and sugar cane have been operated for a number of years, although all require subsidy. The typical capital cost of a wheat-based plant in 1986 producing 130,000 tonnes/year was £50 M. The cost of an equivalent wood-to-ethanol plant producing 76,000 tonnes/year based on acid hydrolysis was £80 M with enzymatic hydrolysis plant estimated at £170 M (the latter feedstocks are much harder to ferment hence the increased capital cost). The most powerful influence on the final price of bioethanol is the cost of the raw material, which represents 40% to 55% of production costs for wood or straw and some 70% for wheat. Overall the cost of producing bioethanol from wheat (one of the easiest feedstocks to ferment) based on feedstock costs of £113/tonne is around 36p/litre. At such a price utilisation of wheat for bioethanol production is not cost-effective and is unlikely to become so in the foreseeable future, despite the availability of adequate conversion technology. Bioethanol would be competing with exrefinery petrol which is currently in the range of 8p to 10p/litre.

Overall, it is difficult to see how bioethanol production from cereals can become economic without substantial subsidy or the value of the competing fossil fuel product increasing very substantially. Similarly, at the existing market price for rape seed of £110/tonne, the cost of untaxed biodiesel (including credits derived from the glycerine and cattle meal by-products of processing) would be approximately 26p/litre. This compares unfavourably with untaxed post-refinery fossil diesel which costs 11p/litre.

F. Resource-Cost Curves

Figures 6 and 7 are for the Maximum Practicable Resource which takes account of the constraints upon the Accessible Resource and how they might change with time. These curves are consistent with the input data used in the energy systems modelling (section H). The use of gasification technology has been assumed throughout (see the Advanced Conversion Technologies): 35% conversion efficiency. The capital cost is assumed to fall from £1.06 M/MW$_e$ in 2005 to £0.883 M/MW$_e$ by 2025. The effects of new conservation

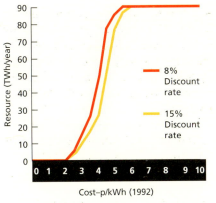
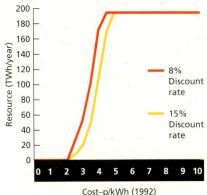

Figure 6: Maximum Practicable Resource–cost curves for energy crops in 2005. In order to minimise anomalies, no grants of any kind have been assumed in these calculations.

Figure 7: Maximum Practicable Resource–cost curves for energy crops in 2025. In order to minimise anomalies, no grants of any kind have been assumed in these calculations.

headlands and field margins have been included to reduce the size of the Accessible Resource by 500,000 ha (2010). Other factors, such as the effect of water availability on crop yield, are currently under research but results are as yet unavailable for application in this analysis. Early results suggest that short rotation coppice may be very tolerant of different soil types and hydrology.

The shape of the resource-cost curve has been generated by making assumptions about the proportion of farmers who will change from food crops to energy crops as the income per hectare from energy crops is increased. Estimates have been made of the surplus land available in 2005 and 2025. The 4 years required to establish a new coppice plantation has been taken into account, but it has been harder to estimate the time delays in winning farmers' and investors' confidence to embark on energy-from-crops projects. Inevitably, there will be some inertia in changing farming practice and a greater appreciation of this effect could be gained during the next tranche of the NFFO.

Projects will be one or two orders of magnitude smaller than conventional power stations and there might be constraints on linking such dispersed generation to the grid, on the other hand, dispersed generation may serve a very useful function in stabilising the grid. This question will be addressed in co-ordination with the other modules. Given the uncertainties, the issue has not been included in this analysis. In addition, the rate at which generating plant could be constructed is difficult to estimate. This is likely to present a barrier to achieving the higher figures for 2005.

A smooth curve has been fitted. This curve starts to rise at very low costs where coppiced wood is available from areas such as those managed for game cover. The curve reaches its asymptotic limit when the electricity price allows the fuel price to be so high that all other options for land surplus to food crop requirements are relatively unattractive.

At this stage, a more sophisticated analysis is inappropriate as there are too many uncertainties in the future. A systematic analysis of farmers' decision making when presented with a range of different land use opportunities is not possible. Also, the amount of surplus land available will depend on Set Aside arrangements which have not been firmed up to give accurate predictions for 2005 and 2025. Although the exact shape of the curve and size of the resource are still a matter for further work, it is undoubtedly true that the resource is potentially extremely large and could be accessible at a realistic electricity price.

G. Constraints and Opportunities

Technological/System

Arable farming is a highly efficient operation in which high levels of mechanisation are routinely employed. Thus adapting existing machinery to harvesting the novel combustible energy crops will reduce the harvesting costs. This is counter to the approach taken initially in the UK and overseas programmes where dedicated harvesters were developed as prototypes. That is the normal approach taken in the forest products industry which had a major influence on the early programmes. Both approaches benefit from involvement of commercial equipment manufacturers. This is often easier to achieve in advance of a market being created if simple adaptation of equipment, rather than new development, is involved. The current UK programme is attracting the support of forage harvester manufactures in trials and machinery modification work but commercial involvement in the development of specific coppice harvesters is difficult to achieve.

With coppice, the high cost of crop establishment appears to be a major constraint to crop adoption. However, as with harvesting, these costs are being reduced by farmers. In this case they often choose not to adopt the forestry practice of installing expensive fencing, choosing instead low cost temporary electric fencing. Additionally, through buying cuttings co-operatively the cost has been reduced from around 20p each to as low as 8p each. This mirrors the situation in Sweden where, with development of a market, cuttings are now available at 3p each. Changes in the Forestry Authority's Woodland Grant Scheme to include the coppice crop should also help to reduce costs. Under the scheme, payments will depend on the area planted, with proportionately larger payments being made for smaller areas. These payments will be phased at 70% in year 1, 20% in year 5 and 10% in year 10 (no maintenance payments will be made). These grants are being reviewed and may change in 1994. In addition, a single 'better land supplement' is payable for land coming out of food production and is currently worth £600/ha. These economics could be bettered by rotational planting. The establishment costs can be reduced by using material cut back from the initial block to produce free cuttings for subsequent planting. In addition, that area of land not planted initially with coppice can remain in food production until required, thus generating further income. These reductions in costs have largely removed the barrier to establishing coppice providing that adequate provision is made under the new Set Aside schemes to allow coppice to be grown. However, this is not the case for the C4 grasses. These are established from rhizomes which are costly to produce and in short supply. This situation will be reversed when the crop is in production as micro propagation techniques can be employed to produce the planting material more cheaply.

Increases in crop yields and biological stability are expected through the adoption of better plant husbandry, clone selection (e.g. greater disease resistance) and breeding. This trend is likely to be substantial and will reduce unit costs in the medium to long term. The commercial development and deployment of improved conversion systems such as gasifiers is now occurring. These increase plant efficiency and, in the future, may be produced with lower capital costs. Other constraints are a lack of planting material and a lack of relevant skills. Each of these needs addressing, the last point through participation of agencies such as the Agricultural Development Advisory Service (ADAS). The major constraint on the adoption of this technology remains the lack of a market for the wood fuel product. A primary goal of any future work will be to establish local fuel outlets simultaneously with crop growth. For the larger scale projects envisaged this is less of a problem since the major product will be electricity, which may be sold initially under NFFO arrangements.

Local generation of electricity, using plant up to perhaps 30 MW$_e$ installed capacity, is clearly possible. Siting may well be rural and remote and will often entail the need for reinforcement of the local line. Such positioning could be useful to the Regional Electricity Company, either obviating the need to install either greater power generation capacity nearby or reducing the need for reinforcement of the line over a wider area. Provided that a sustainable harvesting and storage regime is adopted energy crops, such as arable coppice, should provide a continuous fuel source for both thermal and electrical needs.

Key players for coppice on the production side are the Government Departments who define land use policy and through them ADAS who guide farmers, the Forestry Authority who provide grant assistance, the farming industry (Country Landowners Association, National Farmers Union, Royal Agricultural Society of England and major landowners who define much of current land use) and farming and forestry contractors who provide the link between growers and users. On the utilisation side, the key players are local industry and the Local Authorities. Other interested parties are energy management companies and consultancies with entrepreneurial skills who can pull together both fuel producers and users. For liquid fuel production and use, the oil companies and motor manufacturers will need to be involved. For electricity generation, the large power companies such as National Power, PowerGen and the Regional Electricity Companies will have a major role. Links with sludge disposal and other wastes may also draw in Water Companies and other waste contractors.

For most crops, other than coppice and the C4 grasses, the seasonal nature of growth, harvesting and storage becomes more of a problem. For the perennial crops, winter harvest and subsequent storage has a generally beneficial impact on the crop. Airdrying increases the calorific value of the fuel and makes combustion easier. For wet crops harvesting will generally be in the Autumn and thereafter storage poses a problem. The biomass will begin to deteriorate and decay aerobically, reducing its value as a source of energy. Indeed, unless care is taken over storage, substantial pollution may be caused through the production of high biological oxygen demand liquors, the release of which (to a water course for example) can have catastrophic effects on local wildlife. Hence, the crop must either be used relatively quickly or ensiled in some way. The former route leads to only seasonal operation, the latter to much increased costs of storage.

Institutional
The main driving force behind the interest in energy crop production is CAP reform. The rules which result from these reforms will influence decisions on energy cropping.

As with all renewable energy projects, financing represents a major barrier due in part to the high perception of risk. This is more of a problem with energy crops which will not be produced until a market exists. However, financing the project in the absence of crop production, and thus a firm supply contract, will present an even greater problem.

Convincing farmers that arable coppice is a credible crop to grow is also a problem. This point is currently being addressed through an intensified programme of technology transfer, incorporating field trials. Demonstration of both a fuel supply chain and successful end-use is essential.

Planning permission may well become an issue for larger projects and it will be important to stress the environmental and recreational benefits that are afforded by the growing of crops such as coppice. For example, coppice may be used in the rehabilitation of derelict industrial land providing cover for flora and fauna while allowing public access along designated routes (coppice is a very robust crop once established and difficult to harm). Local Authorities may well wish to take advantage of the screening and sheltering properties of coppice.

There is significant interest from many environmental groups in energy cropping, particularly in coppice. Active participation and peer review are being encouraged in order to maximise the benefits of coppice projects to the local ecology. Coppice is generally viewed as 'green' in the broadest sense. This view needs to be fostered and encouraged to ensure the crop is accepted on as wide a front as possible. Adverse views about its acceptability could restrict the land available and this could seriously affect the development of this technology.

H Prospects Modelling Results

Under the Composite (CSS) scenario, the only contribution from energy crops comes from the NFFO supported projects. By 2025 these plant will have been decommissioned.

With both Low Oil Prices (LOP) and High Oil Prices (HOP) this capacity, whilst available, is not completely utilised due to the high variable costs associated with running the plant. This is borne out by the LOP case in which the utilisation of the NFFO plant is only 15% of that available.

In the Heightened Environmental Concern (HECa) scenario, as much as 78% of the Maximum Practicable Resource is accessed by 2025. Under the HECb scenario the availability of cheap nuclear power at an 8% discount rate entirely excludes energy crops; however, at a 15% discount rate, crops contribute 130 TWh/year, around 68% of the Maximum Practicable Resource.

Earlier contributions under the HEC scenarios mirror this effect. Excluding the situation at 2000, where the contribution is entirely due to NFFO projects, the availability of gasification plant enables the contribution to increase threefold between 2005 and 2010.

Scenario	Discount rate	Contribution, TWh/year			
		2000	2005	2010	2025
HOP	8%	0	0	0.13	0
	15%	0	0.22	0.16	0
	Survey (10%)	0	0.13	0.16	0
CSS	8%	0	0	0.22	0
	15%	0.034	0	0.22	0
	Survey (10%)	0	0	0.22	0
LOP	8%	0	0	0.034	0
	15%	0	0	0.034	0
	Survey (10%)	0	0	0.034	0
HECa	8%	0.22	22	77	130
	15%	0.22	14	39	150
	Survey (10%)	0.22	22	62	140
HECb	8%	0.22	14	14	0
	15%	0.22	0.22	39	130
	Survey (10%)	0.22	21	21	82

Table 8: Potential contribution under all scenarios.

Environmental burden			Incremental contribution (TWh/year)			
	Constraint	Base case scenario	2000	2005	2010	2025
Carbon dioxide	- 5%	CSS	0	0	0	63
	- 5%	HOP	0	-0.13	0	51
Carbon dioxide	-10%	CSS	0	0.16	0	72
	-10%	HOP	0	0.089	0.067	140
Sulphur dioxide	-40%	CSS	0	0.201	0	0
	-40%	HOP	0	0	0.067	0
Nitrogen oxides	-40%	CSS	0	0.22	4.1	130
	-40%	HOP	0	-0.034	0.067	23

Table 9: Incremental contribution under environment constraints.
(TWh/year above base case scenario contribution.)

From table 9 it is clear that, by 2025, the need to reduce carbon dioxide emissions by an additional 5% per decade over the base case scenario results in a significant contribution (33% and 26% of the total Maximum Practicable Resource respectively) under the CSS and HOP scenarios. Where an additional 10% reduction per decade in carbon dioxide emissions over the base case scenario is required, an earlier but modest contribution is achieved by both 2005 and 2010. At 2025, the effect of a HOP scenario is to increase the contribution from crops to 73% of the Maximum Practicable Resource.

Discount rate	Incremental contribution (TWh/year)			
	2000	2005	2010	2025
8%	0.22	0.10	-0.12	0
15%	0.19	0.10	-0.12	56
Survey	0.22	0.10	-0.12	70

Table 10: Incremental contribution under Shifting Sands assumptions (TWh/year above CSS contribution).

II. Programme Review

A. United Kingdom

The use of arable energy crops in the UK is in its infancy. Arable coppice is at the stage where crop trials established by foresters are maturing and sensible predictions about yield can be made. A start is now being made on establishing crops with entrepreneurial farmers and landowners on good quality agricultural land. The industry is at a watershed and markets for wood fuel now need to be found or developed for embryonic farmer co-operatives which are being encouraged in lowland England.

The farming industry is currently in decline but it is the decline in conventional farming for food crops that offers energy cropping a promising future, presenting the farmer with a timely alternative crop. Farmers will play a key role in progressing the energy crop from trials to full scale production showing, hopefully, that the crop can be grown cheaply enough to compete with other fuels.

A number of UK companies have produced pilot plants converting sugar to ethanol by continuous fermentation. No industrial plant has yet been built in the UK but research is being carried out with European Union funding of about £300,000 per year. Areas covered include novel organisms for conversion of cellulose and lignocellulose, production of ethanol from xylose, direct conversion of lignocellulose to ethanol and continuous fermentation of straw.

The Government's Programme
Since the EP55 review the evaluation and development of low cost techniques for producing, harvesting and using forestry residues has continued. In addition, extensive willow and poplar growing trials at various sites around the UK have been established.

The first commercial demonstration of coppice production has been established through the Farm Wood Fuel & Energy Project. This project will not only show the true costs associated with growing this crop, but will also demonstrate what yields might be achieved on good arable land. In association with this project a full environmental impact assessment is being carried out to determine any environmental consequences of growing the crop.

This project has already led to coppice production costs falling through the creation of a market for cuttings as planting material and for machinery associated with the growing of this crop as a farm product. For instance, high fencing costs have been avoided by adopting farming electric fencing as opposed to forestry permanent fencing.

In the early 1980s, assessments of other arable energy crops were carried out. These concluded that such crops would be uneconomic without substantial subsidy or until the price of energy had increased markedly. Currently all energy crops are being reviewed in the light of changes since that time which might affect their viability. In addition, novel high yielding crops such as Miscanthus are being assessed for the first time.

B. Other Countries

Wood fuel, derived from both conventional forestry and (to a lesser extent) arable coppice is currently being used in a number of countries, particularly in Scandinavia and North America. In Scandinavia, the fuel is usually combusted in district heating plants. In North America, it is more usual for the fuel to be used directly by the wood processing industries.

Sweden, in common with Finland and Denmark, conducted much research in the 1970s on harvesting wood fuel from forests. More recently, research has begun to try to increase the wood fuel resource through adoption of coppice systems, mainly willow for climatic reasons. The initial work in Sweden was funded through the National Board for Energy. However, this support is declining with a larger proportion of the expense being borne by the National Farmers Foundation. The Swedish coppice research programme is currently funded at around £1 million per year. Work covers all aspects of cultivation, but latterly has concentrated on development rather than research.

In North America, most research has been directed at fuel wood supply from conventional forestry. However, in eastern Canada some research has been undertaken into coppice fuel wood supply from poplar and willow, with the provinces of Quebec and Ontario funding a willow and poplar breeding programme. In the USA coppice research has been funded by the Department of Energy and administered through the Oak Ridge National Laboratory since the mid-1970s. The level of funding has varied from year to year which has led to lack of continuity.

The UK has access to these foreign national programmes through the Bioenergy Agreement of the International Energy Agency (IEA).

There are several European bioethanol demonstration facilities built by major engineering firms. The only existing plant operating on wheat is the Swedish Biostil plant built by Alfa-Laval producing 20,000 litres per day of ethanol. A number of plants are operating in the USA converting maize to ethanol, the largest being a 600,000 litre per day facility designed by Vogelbusch.

Bioethanol R&D to date has been targeted at improving the energy efficiency of ethanol production, improving glucose and ethanol tolerance of yeast and improving rates of ethanol production.

Biodiesel production is expanding in Europe, although mainly on pilot plant scale. In Austria there are two large production plants: 10,000 tonnes of rape methyl ester (RME) biodiesel per year and 15,000 tonnes RME per year at Aschach and Bruck respectively. In France there is a plant of 20,000 tonnes per year at Compaigne, and further production is planned. Novamont (part of the Ferruzi-Montedison group) have a biodiesel factory at Livorna in Italy, using 60,000 tonnes per year of both rape and soya oil.

R&D into Miscanthus is being carried out in various European countries. A European Union Miscanthus Productivity Network has recently been set up to study productivity and agronomic factors in a range of climate and soil conditions. This includes four sites in the UK. In Germany the Veba Oel company are studying the potential of Miscanthus as a feedstock for gasification, and some pilot combustion facilities, using Miscanthus as a fuel, should be on-line shortly.

Advanced Conversion Technologies

I. Technology Review

A. Status of the Technology

The three principal advanced conversion technologies considered in this module are:

- pyrolysis;
- gasification;
- liquefaction.

These thermochemical processes produce solid, liquid and gaseous intermediate products from biofuels which can be burned to produce heat and/or electricity, or upgraded to substitute fossil fuels. These 'intermediates' can be combusted directly in an engine to produce power without the need to use conventional steam-raising plant. This gives a substantial increase in conversion efficiency at the scales relevant to biofuels (less than 50 MW_e) with capital and operating costs comparable to conventional plant. The economic viability of electricity production from energy crops, forestry wastes and straw is thus significantly improved.

These processes promise to be inherently less polluting than conventional incineration, which is the main technology used for producing electricity from municipal and industrial wastes. By reducing the costs of pollution abatement and offering a more secure disposal route which satisfies stringent pollution legislation, these advanced technologies will enhance the prospects of energy-from-waste schemes. For sewage sludge it is very difficult to make conventional incineration autothermic and then achieve any net power export. Gasification may prove to be the best option for both disposal and power generation.

Pyrolysis
This is the thermal degradation of a feedstock in the absence of an oxidising agent to produce gas, liquid and char. By heating slowly to a low temperature the proportion of char is increased. If the rate of heating is increased, the fraction of liquid and gas produced is raised. At the extreme, this is called flash pyrolysis. The pyrolysis oil may be used to fuel internal combustion engines or it may have value as a chemical feedstock. By this route, it is possible to convert whole crops, including the straw, in the production of liquid biofuels.

Gasification
This is the substoichiometric (i.e. less oxygen than that required for complete combustion) oxidation of a feedstock to produce a gaseous mixture of carbon monoxide, carbon dioxide, methane and hydrogen. For gasifiers which use air as the oxidiser the output gas has a low calorific value, approximately one-fifth that of natural gas, but it can be used directly for heating or in gas engines or turbines.

Coal gasification was used widely in the past to produce town gas but was displaced in the 1960s by natural gas. Now, commercial interest has returned with new plants being commissioned especially in the USA. Biomass gasification also has a long history and was used extensively during World War II for powering vehicles. Simple small scale installations continue to be used around the world. Larger scale gasification of biomass has been in commercial use in Sweden for many years with over 100 MW_{th} installed for use in the paper industry. Within all major national biofuels programmes, gasification has been rising in prominence with a number of countries engaged in large scale demonstrations of the technology. Modern technology has moved a long way from its roots in coal gasification.

Gasification equipment is now available in sizes which allow electricity to be generated in units ranging from 100 kW_e to 30 MW_e using biomass. This covers the range of interest for crops, forestry wastes and straw. Power production from waste is more problematic but it is approaching the pilot scale demonstration stage.

Liquefaction
This is a relatively low temperature process which operates in the liquid phase with high pressure hydrogen injected to liquefy the feedstock. There are a number of technical and engineering requirements which are likely to prevent this process being economically viable, so it will not be considered in any great depth in this module.

Figure 1: Enniskillen gasification plant.

B. Market Status

At present, technical knowledge of the thermal conversion of biofuels is not widespread in the UK. Academic work is concentrated at Aston, Leeds, Strathclyde, Imperial and Warren Spring Laboratories (now incorporated into the National Environmental Technology Centre). The Natural Resources Institute has the most experience of practical trials through their work in the developing countries on small scale equipment.

In terms of commercial plant, Wellman are building on their knowledge of fixed bed coal gasifiers and are contracted to supply equipment for the Border Biofuels 5 MW_e project. In a similar way, the Coal Research Establishment wish to develop their business and are particularly interested in co-firing options involving coal and biomass. Shell strongly supports biomass gasification and have accumulated considerable experience through their studies, but they have taken the decision not to become directly involved in power generation at this time. They have made a major contribution to a large project in Brazil funded by the World Bank. Babcock, John Brown and Foster Wheeler are amongst the major engineering companies which have experience relevant to the manufacture of larger scale gasification plant. In addition, there are a number of small companies offering small gasifiers. There are many small engineering companies capable of manufacturing equipment under licence. For example, Brice Baker is manufacturing the Enniskillen gasifier to a design produced by the University of Louvain, Belgium.

Sweden, Finland, Denmark, Canada and the USA have all benefited from their national biofuels programmes which have maintained support for advanced conversion over many years. In Sweden, the main manufacturer is TPS with the Royal Institute of Technology and Lund University providing academic support. TPS have experience with a 0.5 MW_e pilot plant which has run on biomass for over 1,500 hours. There are plans to build a 6 MW_e plant. Finland's Bioneer and Ahlstrom gasifiers have been successfully commercialised. Fifteen have been sold in sizes up to 35 MW. Denmark has concentrated on studying the advanced conversion of straw. Canada's most interesting area of activity is the Ensyn flash pyrolysis process which is still at the R&D stage. The USA has a wide range of pilot plants based at universities and is about to embark on a major pressurised biomass gasification project in Hawaii.

The driving force in these countries behind the adoption of gasification for converting biomass is the improvement in conversion efficiency. It is commonly believed that this route is more economically attractive than conventional steam plant; it is also regarded as the 'green' approach since the energy balance for crops is improved. In California the installed capacity of conventional plant is 8,000 MW_e and further growth in electricity output is resource-limited. This adds further impetus to improving conversion efficiency.

For initial deployment in the UK, NFFO contracts and strategic grants will be essential since the capital, commissioning and operating costs of initial plant will be too high for them to be viable at the pool price. In the medium to long term, the higher conversion efficiency at comparable or lower capital costs will improve the economics. This should allow the level of NFFO support to be reduced progressively and then removed all together.

Current Status:	
Demonstration	Gasification of crops and agricultural wastes
Development	Flash pyrolysis of crops and agricultural waste
Development	Advanced conversion of municipal and industrial wastes
UK installed capacity	0 MW_{th}
World installed capacity	>200 MW_{th}
UK industry	UK equipment suppliers awaiting opportunity

Table 1: Status of technology.

C. Resource

The resources are described under the relevant biofuels modules – Municipal and General Industrial Wastes, Specialised Industrial Wastes, Agricultural Wastes and Crops.

D. Environmental Aspects

Work on advanced conversion is at too early a stage to be certain of the environmental impacts of the technology. However, for a number of reasons the impact is expected to be lower than that of conventional steam plant.

The location of plant which utilises advanced conversion will be the same as for conventional plant. The visual impact of installations based on these advanced conversion technologies may be slightly less than that for conventional steam cycle plant which may require cooling towers. The plant life expectancy is expected to be similar and decommissioning should not present any special problems.

Improved conversion efficiency means that the carbon dioxide emissions per unit of electricity generated are lower than for conventional plant. The conversion efficiency is expected to increase from 7% for steam plant to 25% for gasification at the 100 kW_e scale and from 20% to 35% at the 15 MW_e scale; carbon dioxide emissions will, therefore, be reduced by over 70% and by over 40% at the two scales respectively. This effect will also reduce other pollutant emissions; in addition they are likely to be further reduced for the following reasons.

- The heavier pollutants will be retained in the ash because of the relatively low reactor temperatures.

- The volatile pollutants, hydrogen chloride and sulphur dioxide, can be removed from the gas output before combustion, giving an advantage over conventional plant where it is necessary to scrub the much larger flue gas flow.

- It is possible that the combustion process will be easier to optimise and so the carbon monoxide and nitrogen oxides emissions will be reduced.

- Preliminary work has indicated that the ash from gasification has good leaching properties allowing safer disposal in landfills.

- As a steam cycle is not used, there is no need for water abstraction and disposal, nor for cooling towers which can cause visual intrusion.

E. Economics At this time it is not possible to present a detailed discounted cash flow analysis of an advanced conversion plant. However, the following statements can be made, based on current knowledge.

- There is a general consensus internationally that gasification will offer the most cost-effective means of utilising biofuels.

- The conversion efficiency gains are widely accepted.

- Data available on prospective projects shows capital costs similar to conventional steam plant at both 5 MW_e and 25 MW_e. The capital costs at the 100 kW_e scale are known to be half or less for gasification/engine sets compared with a small boiler/steam engine. Most observers are optimistic that the capital costs for gasification can be reduced below those for conventional plant. The reduction in the scale of pollution abatement equipment for industrial and municipal wastes has been described earlier. It is almost certain that as steam plant has been refined over many decades the scope for further cost reductions is small. Therefore, all the indications are that the capital costs will be comparable with conventional plant with the strong possibility that they could be lower in the future.

- The operating costs are very difficult to estimate without any experience from commercial plant. There is no reason to suppose they will differ greatly from the costs for steam plant.

The gasification route for biofuel conversion offers prospects of a significant increase in conversion efficiency with commensurate reductions in electricity generating costs.

Co-firing gasified biomass with fossil fuels in a conventional power station offers an opportunity to reduce many of the overheads, reduce capital expenditure and reduce the risks of the project overall.

F. Resource-Cost Curves

Resource-cost curves are not relevant to this module and are covered under the appropriate feedstock modules.

G. Constraints and Opportunities

Technological / System

As regards energy crops and agricultural wastes, the technology is ready for demonstration at both the on-site scale of less than 1 MW_e and at the larger power producer scale of 5 to 30 MW_e. In the past, the most significant technical problem has been that of cleaning up the gas to remove tars and avoid damage to the engine, but this has largely been overcome for these feedstocks by thermal and catalytic cracking. The latter technique has been successfully demonstrated on a pilot plant in Sweden. The capital cost of a one-off demonstration plant is higher than that for the equivalent conventional steam plant, which is generally available 'off the shelf'. These costs will tend to fall into line with those for conventional plant as the market develops.

For industrial and municipal wastes, the problem of gas clean up is not solved at present. The behaviour of the feedstock in a gasifier is not adequately understood. Consequently, more R&D and pilot scale work is required before the technology can move on to demonstration.

Flash pyrolysis offers the potential to make liquid biofuels economically viable, but it is only at the R&D stage and further technical evaluation is necessary.

Co-firing the products of advanced conversion with fossil fuels in a power station is not expected to present any serious technical problems but further economic analysis is needed to quantify the potential benefits.

The problem of alkali metal deposition may present the major technical barrier if gas turbines are to be used, rather than internal combustion engines, at larger scales of operation. If this does become a problem the solution will lie in improving the filtering of the gas before it enters the turbine. If this problem cannot be solved satisfactorily, internal combustion engines can be readily used across the range of capacities of interest. There is no problem in using multiple units for the larger scales.

The level of operator intervention required to run a system is not known at present. For the larger plants it is expected to be comparable with conventional systems. For the smaller on-site generators, it could be difficult to achieve the necessary automation and simple maintenance usually demanded. This will be assessed in the monitoring of demonstration projects.

Institutional

With any novel technology there is a high perceived risk. Demonstration projects and feasibility studies will aim to overcome this barrier. Views on the technology range from high optimism to complete pessimism; it is difficult to form an objective view of the situation with little hard evidence available. The data collected through commercial feasibility studies and the monitoring of demonstration projects will allow more balanced views to be formed.

Work on the gasification of energy crops is growing, stimulated in particular by European Union and World Bank funding. The major national biofuels programmes have identified advanced conversion as the way ahead for energy crops. As liquid biofuel substitutes for fossil fuels rise in prominence the merits of pyrolysis as a potential means of using the whole crop will be more widely recognised. This will create markets in Europe and the developing world for advanced conversion equipment. At present the Scandinavian countries are in the lead. However, UK industry is equipped to manufacture designs under licence, and given time, the technology could be developed and improved by British companies. There are good opportunities for small or medium sized enterprises should small on-site generators prove to be viable.

While the UK technical knowledge is not widespread, there are pockets of expertise which can be built upon to support the future development of a UK advanced conversion industry. The key will be to make maximum use of international experience through, in particular, involvement with IEA activities in this field.

To enable demonstration projects to go ahead, it may be necessary for the early projects not only to secure a NFFO contract but also to receive a grant to offset the capital costs. This funding will be sought where possible through THERMIE and other European Union programmes.

H. Prospects Modelling Results

It has been assumed that from 2000 onwards gasification technology, as well as conventional technologies, will be available to process agricultural wastes and energy crops. Table 2 presents the aggregated gasification generation from these two resources.

Scenario	Discount rate	Contribution, TWh/year			
		2000	2005	2010	2025
HOP	8%	0	0	0	0
	15%	0	0	0	0
	Survey (10%)	0	0	0	0
CSS	8%	0	0	0	0
	15%	0	0	0	0
	Survey (10%)	0	0	0	0
LOP	8%	0	0	0	0
	15%	0	0	0	0
	Survey (10%)	0	0	0	0
HECa	8%	0.07	24	80	130
	15%	0.07	16	42	150
	Survey (10%)	0.07	24	64	140
HECb	8%	0.07	17	17	4.1
	15%	0.07	3	42	130
	Survey (10%)	0.07	23	23	86

Table 2: Potential contribution from agricultural wastes and energy crops via gasification technology under all scenarios. N.B. Within the modules for agricultural wastes and energy crops the total generation is shown; i.e. from both conventional and gasification technology. The uptake shown here should not be added to the uptake shown in the resource modules as this would lead to the gasification contribution being counted twice.

The requirement for gasification technology is clearly strongly influenced by scenario with no deployment estimated under the HOP, CSS and LOP scenarios and very high deployment, in later years, under the HEC scenarios. The bulk of the uptake is from gasification technology being used to process energy crops.

II. Programme Review

A. United Kingdom

Gasification of crops and agricultural wastes is ready for demonstration in the UK. The advanced conversion of municipal and industrial wastes requires more R&D and pilot scale trials before moving to demonstration. Flash pyrolysis is at the research stage. Most importantly, more information is needed to conduct economic analyses of the various options at different scales of operation in comparison with conventional steam plant. Then the benefits of advanced conversion can be quantified, leading to tighter definition of the markets in which it will make the greatest impact. This information can be gathered only by involvement in commercial feasibility studies and demonstration plants in collaboration with industry.

The Government's Programme

Advanced conversion is a new element introduced into the renewables programme following analyses of the cost of generating electricity from energy crops. The opportunity to increase the conversion efficiency presented by advanced conversion offers greater generating cost reductions than are likely to be achieved by reducing fuel costs alone. The Energy from Municipal and Industrial Wastes Programmes have highlighted the impact of pollution legislation on the economic viability of schemes. Advanced conversion has risen in importance because of its long term potential to offer a lower cost route to compliance with ever more stringent legislation.

Since 1977, a number of R&D projects and studies have been supported on the fringe of the other biofuels areas investigating, for example, methanol production from wood, gasification of RDF and liquefaction of wastes. This body of work will be valuable in formulating the future strategy for advanced conversion. It is now clear that the work should be given an individual identity and pushed forward with close UK industrial involvement. The aim will be to encourage deployment of those technologies which offer the best economic advantage for power generation based on biofuels. At an early stage it will be necessary to analyse the economic benefits of different technologies at various scales and for different feedstocks in order to target the programme at the best options.

Under the IEA's Bioenergy Task X, work is being undertaken on gasification and pyrolysis. This provides the UK with an opportunity to benefit from much larger and well established foreign programmes in these areas. The European Union are now placing a greater emphasis on the advanced conversion of biomass through targeted calls for proposals under the THERMIE and JOULE programmes. Liquid biofuels as substitutes for fossil fuels are coming to the fore in Europe and major European Union programmes in this area are likely to be created. The Overseas Development Administration has supported work on small scale plant for applications in the developing world. SERC funding is limited but expertise in this area exists at Aston University, the University of Strathclyde and Imperial College. The work at Warren Spring has been supported by a variety of DTI and DoE projects. The DTI work supported under the Biofuels Programme is becoming a focus for these disparate activities; this should ensure that the UK players reap the benefits of international funding opportunities as they arise.

B. Other Countries

Sweden, Finland, Denmark, Canada and the USA all have major national biofuels programmes which have maintained support for advanced conversion over many years. In Sweden there are plans to build a 6 MW_e plant to be commissioned in 1994. Finland's Bioneer and Ahlstrom gasifiers have been successfully commercialised. 15 have been sold with fuel inputs of up to 35 MW. Denmark has concentrated on studying the advanced conversion of straw. Canada's most interesting area of activity is the Ensyn flash pyrolysis process which is still at the R&D stage. The USA has a wide range of pilot plants based at universities and is about to embark on a major pressurised biomass gasification project in Hawaii. Sweden, Finland and Denmark are all driving towards demonstration and deployment of large plants for biomass conversion. Italy is embarking on demonstration projects converting RDF, but in general municipal and industrial waste projects do not have the same high priority as those for energy crops. This may provide the UK with an opportunity to attack a niche market if R&D is supported in this area now.

Annex 2: RDD&D Requirements and Opportunities

Appraisal of Research, Development, Demonstration and Dissemination

Introduction

The Assessment has been conducted in collaboration with the Appraisal of UK Energy Research, Development, Demonstration and Dissemination (RDD&D) also undertaken by ETSU for the Department of Trade and Industry. The Appraisal covers all energy technologies, both supply (fossil-fuelled, nuclear, renewables) and demand (domestic, industry, transport) as well as various conversion technologies of relevance to the UK. The Assessment of Renewable Energy serves as Volume 3 of the Appraisal.

The Appraisal provides the following for the whole range of energy technologies currently deployed – or potentially deployable – in the UK:

- an estimation of the potential contribution of each technology under a wide range of different future circumstances (scenarios);

- an appraisal of Research, Development, Demonstration and Dissemination in each technology.

The Appraisal methodology has been applied to the renewable technologies. An explanation of the methodology is presented in Volume 8 of the Appraisal. The assessment of their contributions is presented in the technology modules in Annex 1. Each technology module also describes RDD&D currently undertaken in support of the relevant technology, focusing on DTI programmes prior to the current Government review of renewable energy. The RDD&D appraisal presented in this annex consists of the identification of RDD&D requirements and opportunities for the technologies and, where it is applicable, an evaluation of 1992/3 programme expenditure.

RDD&D Requirements and Opportunities

The Appraisal summarises the RDD&D requirements for each technology – that is, the RDD&D which would be necessary if the optimum contribution assessed for each technology is to be attained. By comparing identified requirements with the objectives and content of current programmes, the Appraisal identifies outstanding RDD&D opportunities (i.e. requirements not currently being addressed).

Identification of such opportunities is not in any sense an automatic recommendation for government funding. For the majority of technologies, natural owners of the potential benefits of RDD&D can be identified as oil producers or refiners, electricity generators and plant and equipment manufacturers. It is the natural owners who could most appropriately take up any opportunities highlighted by the Appraisal. For technologies with limited prospects it may be inappropriate for anyone to undertake the identified RDD&D.

In some instances, however, there may be a case for consideration of government support where there is no natural owner, for example:

- where a number of separate organisations have different interests which need to be co-ordinated;

- where the risks may be too high;

- when the potential returns are too long-term for any single natural owner;

- where there is opportunity for the promotion of technology transfer between researchers and manufacturing industry or utility companies to aid commercial deployment.

Programme Evaluation

The Appraisal evaluates current government energy RDD&D programmes in terms of the associated risk and potential payback. The factors chosen to represent risk and payback are drawn from the assessment of each technology's potential contribution. These factors are listed below, together with the measures chosen to rank programmes against each factor.

Factor	Measure
Potential Payback	
Scale	Maximum potential contribution of the technology (Maximum benefits are expressed in 1990 money for consistency with the Appraisal)
Timing	First year in which the technology offers a contribution
Environmental Impact	Incremental performance under carbon dioxide, sulphur dioxide and nitrogen oxides constraints
Export Potential	Star rating (1 to 3) based on expert judgement
Risk	
Technical Risk	Market status of technology (conceptual to deployed)
Market Risk	Scenarios in which the technology offers a contribution
Financial Risk	Scale of RDD&D expenditure
Import Propensity	Star rating (1 to 3) based on expert judgement

The objective of the risk versus payback evaluation is to provide a picture of the current government portfolio, indicating in broad terms the likely cost-effectiveness of each programme in terms of the potential return (national benefit) on government expenditure.

For each renewable energy technology considered, the modules in this annex set out:

- a summary of RDD&D requirements and, by comparison with the 1992/3 programme described in Annex 1, a summary of the RDD&D opportunities identified;

- a standard table illustrating risk and payback evaluations.

Wind Energy

RDD&D Requirements

RDD&D may be identified in a number of areas, which would contribute towards the optimum utilisation of the UK wind energy resource, and the development of appropriate UK capabilities in the domestic and export markets. These areas are set out below.

Research
- To assess in detail the potential and prospects for utilising wind energy (i.e. establish the size, location and accessibility of the practicable onshore resource).

- To undertake a programme of supporting research to obtain new information (e.g. on noise, aerodynamics, materials) which is required to advance the technology to overcome barriers to exploitation.

- To maintain an awareness of the long term potential of offshore generators.

Development
- To pursue, in collaboration with industry, those technical options which appear to offer real scope to reduce the cost of energy.

- To undertake collaborative projects with industry to solve the problems and exploit opportunities exposed by the demonstration projects.

- To identify barriers to exploitation and means to overcome them (e.g. planning guidance, standards, system integration).

Demonstration
- To demonstrate and monitor the improved technology.

- To use available funding to demonstrate periodically the best available technology.

Dissemination
- To disseminate the information obtained by the above activities to all interested parties.

RDD&D Programme

Covered in the wind energy module.

The DTI programme cost £8.7 M in 1992/93.

RDD&D Opportunities

The RDD&D requirements identified above are largely addressed in DTI and/or complementary industrial programmes. However, the following opportunities for further RDD&D have been identified.

- Development of advanced concepts e.g. significantly lighter, more sophisticated turbines.

- Investigation of the system implications of large scale integration, particularly the major resource in Scotland and its integration with England.

- To define better the wind regime by the installation of permanent measuring stations, to complement those of the Meteorological Office, to validate current estimates.

Factor	Measure	Ranking			
Potential Payback					
Scale	Maximum potential contribution of the technology under its best scenario	108 PJ/y			
	Maximum benefit from wind energy	£1,900 M			
Timing	First year in which the technology offers a (new build) contribution in its best scenario	1995			
Environmental impact	Incremental performance in environmental runs:				
	Carbon dioxide	✓✓✓✓			
	Sulphur dioxide	✓			
	Nitrogen oxides	✓✓			
Export potential[1]		**/*			
Risk					
Technical	Market status of technology	Deployed			
Market	Scenarios/discount rate cases in which the technology offers a (new build) contribution	CSS	8		
		HECa	8	15	s
		HECb	8	15	s
		SS	8	15	s
Financial	Scale of RDD&D expenditure (DTI budget for 1992/93)	£8.7 M			
Import propensity[2]		**/*			

Table: Risk versus payback for wind power contribution

[1] Export potential is a medium to high as there is a large potential market available, given a home industry.

[2] Import propensity is also medium to high, depending upon the level of penetration of wind technologies in the UK market.

Hydro Power

RDD&D Requirements

RDD&D may be identified in a number of areas, which would contribute towards the optimum take-up of hydro power in the UK and the development of UK capabilities for the domestic and export markets. These areas are set out below.

Research
- To assess in more detail the potential and prospects for hydro power.

- To quantify and update the hydro power resource in the UK and provide data in order to assess its possible impact on the UK energy scene.

Development
- To undertake feasibility studies.

Demonstration
- To promote the development of hydro power by addressing constraints acting upon the technology (e.g. institutional barriers).

- To make available suitable material and guidance relating to planning matters.

- To demonstrate the use of established technology in the market-place.

- To monitor the costs and performance of appropriate projects, especially with respect to the environmental impact of small hydro schemes.

- To complete programmes on novel low head technology and assess the costs.

Dissemination
- To disseminate information obtained to operators, equipment manufacturers and other interested parties.

RDD&D Programme

Covered in the hydro power technology module.

The DTI programme cost £0.07 M in 1992/93.

RDD&D Opportunities

The RDD&D requirements identified above are largely addressed in DTI and/or complementary industrial programmes. In general future RDD&D opportunities are restricted as hydro technology is close to being fully commercialised. However, the following opportunities for further RDD&D have been identified.

- There are potentially increased opportunities in the dissemination phase of the technology development.

- Further investigation of the opportunities for non-grid connected small hydro in the UK and overseas.

Factor	Measure	Ranking
Potential Payback		
Scale	Maximum potential contribution of the technology under its best scenario	22 PJ/y
	Maximum benefit from hydro power	£360 M
Timing	First year in which the technology offers a (new build) contribution in its best scenario	2000
Environmental impact	Incremental performance in environmental runs: Carbon dioxide Sulphur dioxide Nitrogen oxides	✓ ✓
Export potential[1]		★★
Risk		
Technical	Market status of technology	Deployed
Market	Scenarios/discount rate cases in which the technology offers a (new build) contribution	HOP 8 s CSS 8 s HECa 8 15 s HECb 8 15 s SS 8 15 s
Financial	Scale of RDD&D expenditure (DTI budget for 1992/93)	£0.07 M
Import propensity[2]		★

Table: Risk versus payback for hydro

[1] An export potential ranked as medium is a reflection of the status and prospects of general turbine technology (where there is a lot of overseas competition).

[2] Import propensity is ranked low as the UK is a small market.

Tidal Power

RDD&D Requirements

As the technology has limited prospects in the UK, the current overall aim of the DTI tidal RDD&D programme is to complete the technology assessment and then move to a watching brief. RDD&D may be identified which will help to make this transition, whilst maintaining the option of developing and deploying the technology at a later date. These areas are set out below.

Research

- To resolve the current uncertainty regarding potential environmental impacts, including the lack of impact assessment methodology and tools. This will entail improving the understanding of environmental processes in estuaries and developing techniques for predicting the effects of tidal energy barrages.

- To review the uncertainty inherent in current data on regional effects (i.e. review value of non-energy benefits).

- To investigate the scope for novel financing and other mechanisms for increasing the financial attractiveness of tidal technologies, e.g. by completion of current studies on new construction techniques and an assessment of the tidal stream approach.

- To establish whether reverse economies of scale apply to tidal barrage projects (i.e. see if small schemes are more financially attractive).

Development

- On the above assumptions no development requirements have been identified.

Demonstration

- On the above assumptions no demonstration requirements have been identified.

Dissemination

- On the above assumptions no dissemination requirements have been identified.

RDD&D Programme

Covered in the tidal technology module.

The DTI programme cost £1.7 M in 1992/93.

RDD&D Opportunities

On the assumption of zero commercial potential by current UK criteria (and consistent with maintaining only a watching brief) no RDD&D opportunities have been identified. Were the commercial potential to be demonstrated the primary RDD&D opportunity would be in the provision of a demonstration project, which would generate data relating to engineering topics as well as on environmental matters.

Factor	Measure	Ranking
Potential Payback		
Scale	Maximum potential contribution of the technology under its best scenario	0 PJ/y
	Maximum benefit from tidal power	£0 M
Timing	First year in which the technology offers a (new build) contribution in its best scenario	Not deployed
Environmental impact	Incremental performance in environmental runs: Carbon dioxide Sulphur dioxide Nitrogen oxides	
Export potential[1]		*
Risk		
Technical	Market status of technology	Deployed
Market	Scenarios/discount rate cases in which the technology offers a (new build) contribution	None
Financial	Scale of RDD&D expenditure (DTI budget for 1992/93)	£1.07 M
Import propensity[2]		*

Table: Risk versus payback for tidal

[1] Lack of prospects in the UK, mirror low world-wide interest (bar some interest in countries such as South Korea and Australia) so there is only limited export potential. There may be some scope for consultancy work based on UK expertise in dynamic modelling, shipping effects and environmental matters.

[2] The import propensity is low reflecting the lack of interest elsewhere, and the fact that barrages are fairly low tech devices and hence can be constructed using indigenous capabilities.

Wave Energy

RDD&D Requirements

RDD&D may be identified in a number of areas, which would contribute towards the optimum, but limited, take-up of shoreline wave energy in the UK, and the development of UK capabilities for the domestic and export markets. These areas are set out below.

Research
- To assess in more detail the potential and prospects for shoreline wave energy.

- To continue to study the wave energy resource.

- To carry out initial assessments of new concepts and devices.

Development
- To bring shoreline wave energy technology to a point where it becomes commercially attractive and hence taken up in the marketplace or show that this cannot be done.

- To initiate construction and operation of a demonstration scale prototype shoreline device if earlier results support this action.

- To investigate potential non-technical barriers, e.g. public and commercial acceptance of shoreline devices, regulatory constraints, planning requirements and environmental impacts relating to connections.

Demonstration
- To monitor and optimise the performance of the existing prototype and any future demonstration scale shoreline device(s).

Dissemination
- To disseminate information obtained to potential site operators, equipment manufacturers and other interested parties in order to market and promote the technology to potential industrial partners.

RDD&D Programme

Covered in the wave energy technology module.

The DTI programme cost £0.33 M in 1992/93.

RDD&D Opportunities

As the technology has limited prospects the current overall aim of the DTI wave RDD&D programme is to assess the technology. However, the following opportunities for further RDD&D have been identified.

- To continue to study the wave energy resource.

- To initiate construction and operation of demonstration scale prototype shoreline devices if earlier results support this action.

Factor	Measure	Ranking
Potential Payback		
Scale	Maximum potential contribution of the technology under its best scenario	0.58 PJ/y
	Maximum benefit from wave energy	£4 M
Timing	First year in which the technology offers a (new build) contribution in its best scenario	2000
Environmental impact	Incremental performance in environmental runs: Carbon dioxide Sulphur dioxide Nitrogen oxides	
Export potential[2]		*
Risk		
Technical	Market status of technology	Demonstrated
Market	Scenarios/discount rate cases in which the technology offers a (new build) contribution	HECa 8
Financial	Scale of RDD&D expenditure (DTI budget for 1992/93)	£0.33 M
Import propensity		*

Table: Risk versus payback for shoreline wave[1]

[1] Data for offshore wave technologies have not been presented due to their zero potential contributions.

[2] Export potential and import propensity are both classed as low due to the limited prospects of this technology.

Photovoltaics

RDD&D Requirements

RDD&D may be identified in a number of areas, which would contribute towards the optimum take-up of photovoltaic based technologies in the UK and the development of UK capabilities for the domestic and export markets. These areas are set out below.

Research
- To assess in more detail the potential and prospects for photovoltaic systems integrated into buildings and other small scale systems.

- To investigate the barriers and technical issues associated with deployment of photovoltaic systems.

- To develop and justify cost reduction forecasts.

Development
- To maintain contact with international activities to keep abreast with developments and to obtain benefit from PV installation experiences overseas.

- To develop prototypes and designs of modules and cladding and roofing systems.

- To investigate routes to optimising the integration of PV systems into buildings.

- To assess the environmental impact of significant deployment of PV systems in the urban environment.

Demonstration
- To obtain system performance data and information on technical and non-technical barriers to deployment.

Dissemination
- To disseminate information obtained to operators, equipment manufacturers and other interested parties.

RDD&D Programme

Covered in the photovoltaics module.

The DTI programme cost £0.20 M in 1992/93.

RDD&D Opportunities

The RDD&D requirements identified above are largely addressed in DTI and/or complementary industrial programmes. However, the following opportunities for further RDD&D have been identified.

- Fundamental R&D on basic semiconductor technologies, e.g. in new materials.

- RDD&D considering application of PV in remote power applications. Such opportunities exist as the application of PV in niche markets both in the developed and developing countries may give opportunities for technologies etc. which do not look attractive for large scale commercial deployment. There is therefore an opportunity for RDD&D aimed at such export markets.

- Power conditioning and control including inverter design and intelligent metering.

- Development of DC appliances; e.g. lighting systems.

- Development of PV applications for bulk supply and support to weak grids.

- Improved battery and other storage technologies.

- Reduce unit costs, e.g. mass production manufacturing techniques.

Factor	Measure	Ranking			
Potential Payback					
Scale[1]	Maximum potential contribution of the technology under its best scenario	26 PJ/y			
	Maximum benefit from photovoltaics	£170 M			
Timing	First year in which the technology offers a (new build) contribution in its best scenario	2000			
Environmental impact	Incremental performance in environmental runs:				
	Carbon dioxide	✓ ✓ ✓ ✓			
	Sulphur dioxide				
	Nitrogen oxides	✓ ✓			
Export potential[2]		***			
Risk					
Technical	Market status of technology	Demonstrated			
Market	Scenarios/discount rate cases in which the technology offers a (new build) contribution	HOP	8	s	
		CSS	8	15	s
		LOP	8		
		HECa	8	15	s
		HECb	8	15	s
Financial	Scale of RDD&D expenditure (DTI budget for 1992)	£0.20 M			
Import propensity		***			

Table: Risk versus payback for PV

[1] Reported potential contributions, apart from in the HEC scenarios, only occur towards the end of the study period.

[2] Export potential is classed as high due to existence of current capacity and expertise within the UK and the large overseas market potentially available. Import propensity is high due to the existence and probable significant development of overseas capacity.

Photoconversion

RDD&D Requirements   The technologies associated with photoconversion are at an early stage in their development. There are as yet no indications that they will appear to be commercially viable in application. Until such time as commercial viability looks feasible a watching brief on the technology is all that is required of the DTI photoconversion RDD&D programme. RDD&D may be identified which will help to make this transition, whilst maintaining the option of developing and deploying the technology at a later date. These areas are set out below.

Research
- To identify and assess the key photoconversion technologies in terms of their performance, costs and resource size.

- To determine and specify the primary issues associated with the potential development and deployment of these technologies, notably in information relating to how systems scale up from laboratory processes.

- To maintain awareness of developments in key photoconversion technologies in academic and industrial areas.

- To refine estimates of performance characteristics and economics of appropriate photoconversion systems.

- To assess the environmental impacts of photoconversion technologies.

- To maintain watching brief on other photoconversion technologies.

Development
- On the above assumptions no development requirements have been identified.

Demonstration
- On the above assumptions no demonstration requirements have been identified.

Dissemination
- On the above assumptions no dissemination requirements have been identified.

RDD&D Programme   Covered in the photoconversion module.

The DTI programme cost £0 M in 1992/93.

RDD&D Opportunities   The following RDD&D opportunities have been identified.

- Development of extended electrolyte lifetimes.

- Identification of suitable anaerobic organisms for maximum hydrogen production and suitable separation systems.

- Identification of suitable growth media for biological systems.

- Development of improved cell semi-conductor surfaces and containment structures.

Factor	Measure	Ranking
Potential Payback		
Scale[1]	Maximum potential contribution of the technology under its best scenario	Not modelled
	Maximum benefit of photoconversion	Not modelled
Timing	First year in which the technology offers a (new build) contribution in its best scenario	Not modelled
Environmental impact	Incremental performance in environmental runs:	
	Carbon dioxide	Not modelled
	Sulphur dioxide	Not modelled
	Nitrogen oxides	Not modelled
Export potential[2]		***
Risk		
Technical	Market status of technology	Conceivable
Market	Scenarios/discount rate cases in which the technology offers a (new build) contribution	Not modelled
Financial	Scale of RDD&D expenditure (DTI budget in 1992)	£0 M
Import propensity		**

Table: Risk versus payback for photoconversion

[1] The technology was not subjected to the modelling process as insufficient data was available because it has not yet been assessed. Consequently, there are no 'scale' and 'market', which are results from the modelling exercise.

[2] Although the export potential is considered to be high, the available data indicates that the technology will not begin to look promising for some time yet.

Active Solar including Solar Aided District Heating (SADHS) and Thermal Solar Power

RDD&D Requirements

RDD&D may be identified in a number of areas, which would contribute towards the optimum take-up of active solar and solar aided district heating (SADHS) technologies in the UK, and the development of UK capabilities for the domestic and export markets. These areas are set out below: Solar thermal power has been assessed as inappropriate for the UK climate so no R&D requirements have been identified, other than watching brief.

Research
- To assess in more detail the prospects for SADHS in the UK.

- To determine the health risk associated with active solar water heating.

- To maintain a watching brief on the development of the markets and prospects for active solar heating, thermal solar power and SADHS, and analyse the likely future trends in these markets.

Development
- No development requirements have been identified.

Demonstration
- To provide information and guidance in order to exploit fully the potential market (by eroding public, institutional and industrial barriers caused by the lack of knowledge and confidence in active solar technology) and increase public confidence in the technology.

Dissemination
- No dissemination requirements have been identified.

RDD&D Programme

Covered in the active solar heating technology module.

The DTI programme cost £0 M in 1992/93.

RDD&D Opportunities

The RDD&D requirements identified above are largely addressed in DTI and/or complementary industrial programmes. However, the following opportunities for further RDD&D have been identified.

- To develop further integrated collector/storage systems and to improve their capacity and feasibility.

- To develop further the technology of solar-aided district heating systems and storage mechanisms to make large scale active solar schemes more attractive.

- Performance evaluation of advanced collector designs.

- Development of advanced low cost selective surfaces for collectors.

- Development of thermo-syphon designs with indoor storage adapted to UK houses.

- Development of methods to characterise complete solar water heating systems.

Factor	Measure	Ranking
Potential Payback		
Scale	Maximum potential contribution of the technology under its best scenario	9 PJ/y
	Maximum benefit from active solar design	£75 M
Timing	First year in which the technology offers a (new build) contribution in its best scenario	2015
Environmental impact	Incremental performance in environmental runs: Carbon dioxide Sulphur dioxide Nitrogen oxides	
Export potential[1]		***
Risk		
Technical	Market status of technology	Deployed
Market	Scenarios/discount rate cases in which the technology offers a contribution	CSS 8 HECa 8 HECb 8
Financial	Scale of RDD&D expenditure (DTI budget for 1992/93)	£0 M
Import propensity[2]		**

Table: Risk versus payback for active solar heating

[1]   Export potential is high due to existing UK capacity.

[2]   Import propensity is medium due to presence of international competition (and lower than export as competition will probably be looking at their home markets and high solar potential markets as opposed to the UK).

Passive Solar Design

RDD&D Requirements

RDD&D may be identified in a number of areas, which would contribute towards the optimum take-up of passive solar design in the UK and the development of UK capabilities for the domestic and export markets. These areas are set out below.

Research
- To assess in more detail the main passive solar design measures including their energy and non-energy benefits.

- To assess the full energy benefits of "simple" passive solar design for housing and non-domestic properties.

Development
- To develop design guidance for identified key building types (notably offices, schools and factories).

- To explore the potential for, and then develop appropriate design guidance relating to, the uptake of new glazing technologies and their impact on daylighting design measures.

Demonstration
- To demonstrate and monitor the application of a range of these measures, including daylighting, ventilation and heating, on the various building types.

Dissemination
- To disseminate information obtained to the principal parties in the key building sectors, e.g. property developers, architects, services engineers and lighting engineers

- To increase market take-up of passive solar design by means of marketing and design advice.

RDD&D Programme

Covered in the passive solar design technology module.

The DTI programme cost £2.1 M in 1992/93.

RDD&D Opportunities

The RDD&D requirements identified above are either directly addressed in the DTI passive solar design programme or are being undertaken by industrial concerns. However, the following RDD&D opportunities have been identified.

- Development of improved air collector systems, e.g. space collector systems.

- Development of cost-effective heat storage systems, e.g. phase change materials.

- Development of low cost transparent insulation systems for retrofit applications.

- Development of advanced low energy house designs.

- Assessment of energy savings and subjective reactions to new forms of lighting control.

- Assessment of occupant requirements and maintenance needs of daylighting systems.

Factor	Measure	Ranking		
Potential Payback				
Scale	Maximum potential contribution of the technology under its best scenario[1]	21 PJ/y		
	Maximum benefit from passive solar design	£630 M		
Timing	First year in which the technology offers a (new build) contribution in its best scenario	1995		
Environmental impact	Incremental performance in environmental runs:			
	Carbon dioxide			
	Sulphur dioxide			
	Nitrogen oxides			
Export potential[2]		**		
Risk				
Technical	Market status of technology	Demonstrated		
Market	Scenarios/discount rate cases in which the technology offers a contribution	HOP	8 15	s
		CSS	8 15	s
		LOP	8 15	s
		HECa	8 15	s
		HECb	8 15	s
Financial[3]	Scale of RDD&D expenditure (DTI budget for 1992/93)	£2.1 M		
Import propensity[4]		*		

Table: Risk versus payback for passive solar design

[1] The figures for maximum contribution and benefit include new build PSD and a broad estimate of possible refurbishment.

[2] Export potential is medium due to existing UK capacity (knowledge).

[3] The 1992/93 programme cost of £2.1 M comprised £1.53 M on R,D&D and £0.57 M on the Energy Design Advice Scheme (EDAS).

[4] Import propensity is low since the technology is largely non-hardware based, and design issues can be country specific.

Geothermal HDR

RDD&D Requirements

As the technology has poor prospects the current overall aim of the DTI geothermal HDR RDD&D programme is to assess the technology and then move to a watching brief (based on the conclusion of the 1990 review of this technology that very substantial uncertainties remain about the practicability of extracting energy from HDR within the UK). However, RDD&D may be identified which will help to make this transition, whilst maintaining the option of developing and deploying the technology at a later date. These areas are set out below.

Research
- To maintain a watching brief within the UK and Europe to maintain the option of developing and deploying the technology should there be substantial changes in the technology.

- To ensure that UK expertise is exploited within European HDR programmes.

Development
- To continue addressing the technical uncertainties and study further the opportunities for cost reductions, offered by developments in power generation, drilling and downhole engineering.

Demonstration
- On the above assumptions no demonstration requirements have been identified.

Dissemination
- Promotional activities will be restricted to the provision of information to the HDR scientific community.

RDD&D Programme

Covered in the geothermal HDR technology module.

The DTI programme cost £1.1 M in 1992/93.

RDD&D Opportunities

On the assumption of zero commercial potential (and consistent with maintaining only a watching brief) no RDD&D opportunities have been identified.

Factor	Measure	Ranking
Potential Payback		
Scale	Maximum potential contribution of the technology under its best scenario	0 PJ/y
	Maximum benefit from geothermal HDR	£0 M
Timing	First year in which the technology offers a (new build) contribution in its best scenario	Not deployed
Environmental impact	Incremental performance in environmental runs: Carbon dioxide Sulphur dioxide Nitrogen oxides	
Export potential[1]		*
Risk		
Technical	Market status of technology	Conceivable
Market	Scenarios/discount rate cases in which the technology offers a (new build) contribution	None
Financial	Scale of RDD&D expenditure (DTI budget for 1992/93)	£1.1 M
Import propensity		*

Table: Risk versus payback for geothermal HDR

[1] Export potential and import propensity are both classed as low to reflect the limited prospects for technology.

Geothermal Aquifers

RDD&D Requirements

As the technology has poor prospects the current overall aim of the DTI geothermal aquifers RDD&D programme is to assess the technology and then move to a watching brief. The aquifer resource in the UK is limited and unlikely to be cost-effective in the short to medium term. However, some RDD&D is necessary during the transition period, set out below.

Research
- To assess in more detail the UK aquifer resource and seal any open boreholes if commercial exploitation cannot be effected.

- To carry out a detailed assessment of the economics of energy recovery from geothermal aquifers in the UK.

Development
- On the above assumptions no development requirements have been identified.

Demonstration
- On the above assumptions no demonstration requirements have been identified.

Dissemination
- On the above assumptions no dissemination requirements have been identified.

RDD&D Programme

Covered in the geothermal aquifers technology module.

The DTI programme cost £0.13 M in 1992/93.

RDD&D Opportunities

On the assumption of zero commercial potential (and consistent with maintaining only a watching brief) no RDD&D opportunities have been identified.

Factor	Measure	Ranking
Potential Payback		
Scale	Maximum potential contribution of the technology under its best scenario[1]	0 PJ/y
	Maximum benefit from geothermal aquifers	£0 M
Timing	First year in which the technology offers a (new build) contribution in its best scenario	Not deployed
Environmental impact	Incremental performance in environmental runs: Carbon dioxide Sulphur dioxide Nitrogen oxides	
Export potential[2]		*
Risk		
Technical	Market status of technology	Demonstrated
Market	Scenarios/discount rate cases in which the technology offers a (new build) contribution	None
Financial	Scale of RDD&D expenditure (DTI budget for 1992/93)	£0.13 M
Import propensity		*

Table: Risk versus payback for hydro

[1] A potential contribution equivalent to the 16 GWh/y output from the Southampton scheme was not modelled and hence is not reported.

[2] Export and import potentials are low as there are limited prospects for the technology.

Municipal and Industrial Wastes

RDD&D Requirements   RDD&D may be identified in a number of areas, which would contribute towards the optimum take-up of technologies in the UK and the development of UK capabilities for the domestic and export markets. These areas are set out below.

Research
- To assess in more detail the potential and prospects of the range of waste to energy options (from both 'wet' and 'dry' feed stocks).

- To assess the available potential and, where appropriate, generate resource maps.

- To establish the environmental impacts of waste combustion and compare with other disposal/treatment options.

- To develop an understanding of pollutant formation and destruction mechanisms and of abatement systems performance.

- To investigate alternative methods of treating/disposing of residues.

Development
- To establish and assess the costs/performance/emission levels relevant to the available options.

Demonstration
- To establish that energy recovery is a legitimate part of good resource recovery practice and aid the demonstration of a range of energy recovery technologies.

Dissemination
- To address barriers to implementation caused by current public perception of waste to energy plants.

- To develop and provide information and guidance to developers, local authorities and others involved in waste to energy projects.

- To promote and publicise successful examples of industrial waste to energy schemes.

RDD&D Programme   Covered in the MSW and industrial wastes modules.

The DTI MSW programme cost £0.66 M in 1992/93. The DTI Industrial Wastes programme cost £0.20 M in 1992/93.

RDD&D Opportunities   The RDD&D requirements identified above are either directly addressed in the DTI programme or are being undertaken by UK facility operators and/or equipment manufacturers. However the following opportunities for further RDD&D have been identified.

- To support 'best practice' type schemes to demonstrate state-of-the-art options.

- To undertake pilot or full scale demonstration trials on fluidised bed combustion systems.

Factor	Measure	Ranking
Potential Payback		
Scale	Maximum potential contribution of the technology under its best scenario	22 PJ/y
	Maximum benefit of MSW	£660 M
Timing	First year in which the technology offers a (new build) contribution in its best scenario	2000
Environmental impact	Incremental performance in environmental runs:	
	Carbon dioxide	*X X X X*
	Sulphur dioxide	
	Nitrogen oxides	*X X*
Export potential[1]		*
Risk		
Technical	Market status of technology	Deployed
Market	Scenarios/discount rate cases in which the technology offers a (new build) contribution	HOP 8 CSS 8 15 s LOP 8 HECa 8 s HECb 8 SS 8 15 s
Financial	Scale of RDD&D expenditure (DTI budget for 1992/93)	
	Municipal Solid Waste	£0.66 M
	Industrial Waste	£0.20 M
Import propensity		***

Table: Risk versus payback for MSW

[1] The UK is totally dependent upon importing strategic parts of MSW technology, therefore import propensity is high and export potential low.

Landfill Gas

RDD&D Requirements

RDD&D may be identified in a number of areas, which would contribute towards the optimum take-up of landfill gas (LFG) technology in the UK and the development of UK capabilities for the domestic and export markets. These areas are set out below.

Research
- To assess in more detail the potential and prospects for LFG utilisation.

- To increase confidence in estimates of gas yields from specific sites.

Development
- To reduce costs and improve efficiency of current LFG technology.

- To optimise gas yields and collection efficiency.

- To examine the potential benefits of decision analysis techniques applied to the evaluation of the energy and environment options related to landfill and other waste disposal options.

Demonstration
- To collect and collate information on technical and economic performance of currently existing LFG projects.

- To add data on the technical and economic performance of new LFG projects to the above data.

- To review the institutional and other barriers to LFG exploitation and identify ways to overcome them.

Dissemination
- To disseminate information obtained to site operators, equipment manufacturers, regulatory bodies, financiers and other interested parties.

RDD&D Programme

Covered in the LFG technology module.

The DTI programme cost £0.61 M in 1992/93.

RDD&D Opportunities

The RDD&D requirements identified above are either directly addressed in the DTI LFG programme or are being undertaken by UK site operators and/or equipment manufacturers. However the following opportunity for further RDD&D has been identified.

- There may be opportunities to increase the exploitable reserves of landfill gas by implementing advanced landfilling and site management techniques.

Factor	Measure	Ranking
Potential Payback		
Scale	Maximum potential contribution of the technology under its best scenario	31 PJ/y
	Maximum benefit of LFG	£1,074 M
Timing	First year in which the technology offers a (new build) contribution in its best scenario	1995
Environmental impact	Incremental performance in environmental runs:	
	Carbon dioxide	✓ ✓ ✓ ✓
	Sulphur dioxide	
	Nitrogen oxides	✓ ✓
Export potential[1]		*
Risk		
Technical	Market status of technology	Deployed
Market	Scenarios/discount rate cases in which the technology offers a (new build) contribution	HECa 8 15 s HECb 8 15 s SS 8 15 s
Financial	Scale of RDD&D expenditure (DTI budget for 1992/93)	£0.61 M
Import propensity[2]		*

Table: Risk versus payback for LFG

[1] Landfill gas exploitation in the UK is dominated by a developing group of domestic companies with expertise including consultancy, manufacture and installation. The technology is widely available overseas, and so export potential may be limited. UK equipment and expertise compares well with that of overseas competitors so import propensity is also judged to be low.

Agricultural and Forestry Wastes

RDD&D Requirements
RDD&D may be identified in a number of areas, which would contribute towards the optimum take-up of agricultural waste based technologies in the UK, and the development of UK capabilities for the domestic and export markets. These areas are set out below.

Research
- To assess in more detail the potential and prospects of agricultural and forestry wastes as fuels.

- To maintain a watching brief on developments in anaerobic digestion processes as a means of producing energy from wet wastes or other green agricultural feedstocks.

Development
- To develop improved equipment to mechanise the supply of agricultural wastes as fuels to a range of end users.

- To determine the most cost-effective storage methods for the fuels, especially those which are harvested seasonally.

- To define supply strategies for the most cost-effective means by which fuel may be harvested, stored, transported and delivered to the customer.

Demonstration
- To complete the assessment of potential non-technical barriers, especially those associated with: environmental impacts; planning; gaining access to grid connections; and achieving long term fuel supply contracts.

- To demonstrate the use of agricultural wastes as fuel for the generation of power and heat.

Dissemination
- To disseminate information obtained in demonstration projects to site operators, equipment manufacturers and other interested parties.

- To use results and experience gained to achieve technology transfer to commercial scale projects.

RDD&D Programme
Covered in the agricultural and forestry wastes technology module.

The DTI programme cost £0.24 M in 1992/93.

RDD&D Opportunities
The RDD&D requirements identified above are either directly addressed in the DTI programme or are being undertaken by relevant UK industrial concerns. However the following opportunity for further RDD&D has been identified.

- Following successful demonstration, the need for specialised harvesting and fuel processing equipment may be identified.

Factor	Measure	Ranking
Potential Payback		
Scale	Maximum potential contribution of the technology under its best scenario	23 PJ/y
	Maximum benefit from agricultural and forestry wastes	£660 M
Timing	First year in which the technology offers a (new build) contribution in its best scenario	1995
Environmental impact	Incremental performance in environmental runs: Carbon dioxide Sulphur dioxide Nitrogen oxides	 ✓ ✓ ✗ ✗
Export potential[1]		***
Risk		
Technical	Market status of technology	Demonstrated
Market	Scenarios/discount rate cases in which the technology offers a (new build) contribution	HECa 8 15 s HECb 8 15 s SS 15 s
Financial	Scale of RDD&D expenditure (DTI budget for 1992/93)	£0.24 M
Import propensity		***

Table: Risk versus payback for agricultural and forestry wastes

[1] There is an existing international market in incineration technologies, with an existing UK presence; therefore both export potential and import propensity are high.

Energy Crops

RDD&D Requirements

RDD&D may be identified in a number of areas, which would contribute towards the optimum take-up of energy crop based technologies in the UK, and the development of UK capabilities for the domestic and export markets. These areas are set out below.

Research
- To assess in more detail the potential and prospects for energy crops.

- To initiate clonal selection & improvement programmes to achieve increased yields.
- To initiate studies to address potential disease problems and their management.

Development
- To develop supply strategies addressing the most effective means of producing, harvesting, storing and transporting energy crops.

- To complete the development of robust production, harvesting, processing and storage methods.

Demonstration
- To assess the potential impact of the following non-technical barriers: the environmental impacts of energy crops; landscape and planning issues; gaining access to grid connections; and achieving long term fuel supply contracts.

- To initiate energy crop demonstration in association with the farming industry.

Dissemination
- To market and promote the results of the RD&D programme to farmers, landowners, potential users, equipment suppliers and financiers through a mix of marketing and promotional activities designed to address the needs of key target audiences.

RDD&D Programme

Covered in the energy crops technology module.

The DTI programme cost £1.2 M in 1992/93.

RDD&D Opportunities

The RDD&D requirements identified above are either directly addressed in the DTI energy crop programme and other interested parties. No gaps have been identified as presenting significant new RDD&D opportunities. Whilst no new opportunities have been identified opportunities exist to expand the current programmes on crop breeding and selection as well as on dissemination programmes.

Following the successful demonstration of energy crop production, opportunities in the following areas may be identified:

- Development of improved crop varieties.

- Development of improved farming methods including: crop establishment techniques, crop husbandry techniques and improved harvesting and fuel processing equipment.

Factor	Measure	Ranking		
Potential Payback				
Scale[1]	Maximum potential contribution of the technology under its best scenario	540 PJ/y		
	Maximum benefit from energy crops	£2,000 M		
Timing	First year in which the technology offers a (new build) contribution in its best scenario	2005		
Environmental impact	Incremental performance in environmental runs:			
	Carbon dioxide	✓ ✓ ✓ ✓		
	Sulphur dioxide			
	Nitrogen oxides	✓ ✓		
Export potential		***		
Risk				
Technical	Market status of technology	Prospective		
Market	Scenarios/discount rate cases in which the technology offers a (new build) contribution	HECa	8	15 s
		HECb	8	15 s
		SS		15 s
Financial	Scale of RDD&D expenditure (DTI budget for 1992/93)	£1.2 M		
Import propensity		**		

Table: Risk versus payback for energy crops

[1] The maximum contribution figure is based on utilisation of energy crops for electricity production only (i.e. it does not include energy crop use for hydrogen production).

Advanced Conversion Technologies

RDD&D Requirements

RDD&D may be identified in a number of areas, which would contribute towards the optimum take-up of advanced conversion technologies in the UK and the development of UK capabilities for the domestic and export markets. These areas are set out below.

Research
- To assess in more detail the potential and prospects of gasification, pyrolysis and liquefaction technologies.

- To assess the different technical options for converting feedstocks into suitable fuels.

- To investigate the options for producing and upgrading liquid biofuels - e.g. by conversion to methanol, catalytic reforming, pyrolysis and liquefaction - and for their utilisation.

- To maintain a watching brief on relevant innovations exploring opportunities which have the potential to advance the technology significantly.

Development
- To identify and address the critical barriers that may hinder future deployment

- To carry out economic analyses as necessary, especially with regard to assessing the implications of cost reduction opportunities.

- To survey the markets for the chars and liquid fuels that may be by-products of these technologies.

Demonstration
- To undertake feasibility studies for new (commercial) plant deployment.

- To monitor the characteristics and performance of both UK and overseas plants.

Dissemination
- To market and promote results of the RD&D programme to equipment suppliers, potential users and financiers through a mix of promotional tools such as fact sheets and summaries (possibly by integration with the waste and energy crop programmes).

RDD&D Programme

Covered in the advanced conversion technology module.

The DTI programme cost £0.20 M in 1992/93.

RDD&D Opportunities

The RDD&D requirements identified above are largely addressed in DTI and/or complementary industrial programmes. However, the following opportunities for further RDD&D have been identified.

- Establishment of a centre of expertise on advanced conversion technologies. This centre could link academic, government and industrial programmes.

- Provision of strategic grants for demonstration projects (at 5 MW$_e$ or greater).

- Development programme for automating the operation of gasifiers.

- Improvement of the UK's pilot plant facilities.

- Development of applications to utilise fuels prepared using advanced conversion techniques.

Factor	Measure	Ranking
Potential Payback		
Scale[1]	Maximum potential contribution of the technology under its best scenario	560 PJ/y
	Maximum benefit from advanced conversion technologies	£1,900 M
Timing	First year in which the technology offers a (new build) contribution in its best scenario	2000
Environmental impact	Incremental performance in environmental runs:	
	Carbon dioxide	✓✓✓✓
	Sulphur dioxide	
	Nitrogen oxides	✓✓
Export potential[1]		***
Risk		
Technical	Market status of technology	Demonstrated
Market	Scenarios/discount rate cases in which the technology offers a (new build) contribution	HECa 8 15 s HECb 8 15 s SS 15 s
Financial	Scale of RDD&D expenditure (DTI budget in 1992)	£0.20 M
Import propensity[2]		***

Table: Risk versus payback for advanced conversion technologies

[1] The contribution is based on thermal processing of straw and energy crops for electricity production only. In practice the technology will have applications to many more fuels and as such larger potential contributions and increased maximum benifit are feasible.

Notes

Money   All monies are in 1992 values unless otherwise stated.

Terms used

J	Joule
W	Watt (1 Joule per second)
Wh	Watt-hour
p/kWh	pence per kilowatt hour
Wh/year	Watt-hours per year
Wp	Watt peak (A term used in photovoltaic technology denoting the power delivered under standard conditions, which for flat plate cells are solar radiation of 1,000W/sq.m, a cell temperature of 25°C and an air mass of 1.5).
ha	Hectare (10,000 sq.m)

Unit Prefixes   These scaling factors may be combined with any of the above units.

kilo (k)	= 1,000	= 1×10^3
Mega (M)	= 1,000,000	= 1×10^6
Giga (G)	= 1,000,000,000	= 1×10^9
Tera (T)	= 1,000,000,000,000	= 1×10^{12}
Peta (P)	= 1,000,000,000,000,000	= 1×10^{15}

Electrical Units   A kilowatt-hour (kWh) is the energy expended when a rate of 1,000 watts is maintained for an hour. For instance a typical one bar electric fire would use electricity at the rate of one kilowatt-hour every hour and its cost could be expressed in pence for every kilowatt-hour i.e. p/kWh.

If 1,000,000,000 kWh (1TWh) is supplied or consumed during one year, it is expressed as 1 TWh/year.

Declared Net Capacity (DNC)   Broadly, the DNC for intermittent renewables is the equivalent capacity of base-load plant that would produce the same average annual energy output.

Resource Assessment   In this work, which assesses the whole of the UK, it was not possible to undertake a detailed analysis of the complex system integration and operation issues which would result from the large-scale deployment of renewables. However, it is certain that these factors would constrain the resource and increase its cost but to what degree is currently unknown.

Further Reading:

General Reading

Renewable Energy in the UK: The Way Forward. Energy Paper Number 55. HMSO ISBN 0 11 412905 3 (1988).

Renewable Energy Advisory Group: Report to the President of the Board of Trade. Energy Paper Number 60, November 1992, Department of Trade and Industry. HMSO ISBN 0 11 414287 4.

Renewable Energy: Sources for Fuels and Electricity. Ed. Johansson, TB et al. Island Press. ISBN 1 55963 139 2 (1993).

Planning for Renewable Energy in Devon. ETSU, Devon County Council and West Devon Borough Council (1993).

Prospects for Renewable Energy in Northern Ireland. ETSU, the Department of Economic Development and Northern Ireland Electricity plc. (1993).

Renewable Sources of Electricity in the SWEB Area: Future Prospects. A joint study by South Western Electricity plc (SWEB) and ETSU (1993).

Prospects for Renewable Energy in the NORWEB Area: ETSU and NORWEB. ISBN 0 7058 1504 8 (1989).

Sustainable Development: The UK Strategy. HMSO 010 124 262X (1994).

Renewable Energy Sources, Report Number 22: Watt Committee on Energy, Elsevier Applied Science. ISBN 1 85166 500 5 (1990).

This Common Inheritance: Britain's Environmental Strategy: Government White Paper. HMSO ISBN 0 10 112002 8 (September 1990).

A Summary of the Main Legislation Affecting the Deployment of Renewable Energy in the United Kingdom: Proceedings of the EURES Conference 3-5 June 1992, Seville, Spain. (In press).

Planning Policy Guidance: Renewable Energy. Department of the Environment & Welsh Office, PPG 22. HMSO. ISBN 0 11 752756 4 (1993).

Wind Power

Wind Energy and the Environment. Ed: Swift-Hook, DT IEE Energy Series 4. Published by Peter Perigrinus on behalf of the IEE. ISBN 0 86341 176 2 (1989).

Wind Energy Conversion Systems. Freris, LL. Published by Prentice Hall. ISBN 0 13 960527 4 (1990).

The Commercial Development of Wind power in Europe: Status and Perspectives. Garrad, AD in European Wind Energy Association Special Topic Conference – The Potential of Wind Farms at Herning. Denmark, September 1992.

Computer modelling of the UK Wind Energy Resource: Final Overview Report. ETSU WN 7055.

Attitudes Towards Wind: a Survey of Opinion in Cornwall and Devon. ETSU W 13 00354 038 REP.

Hydro Power   Small-Scale Hydro-Electric Generation Potential in the UK: Salford Civil Engineering Limited, (P1, 2 and 3). ETSU SSH 4063 (1989).

Small-Scale Hydro Power Study of Non-Technical Barriers. ETSU SSH 4063 (1989).

Civil Engineering Hydraulics. Featherstone, RE and Nalluri, C. BSP Professional Books, Oxford (1988).

Tidal Power   The UK Potential for Tidal Energy from Small Estuaries. ETSU TID 4048 P1 (1989).

Tidal Power. Baker, AC. Institution of Electrical Engineers. Peregrinus. ISBN 08634 11894 (1991).

The Severn Barrage Project General Report. Energy Paper No 57. HMSO. (1989).

Severn Barrage Project Detail Report. ETSU TID 4060 Vols. I to V (1989).

Severn Barrage Project Report on Responses and Consultation following Publication of Energy Paper 57. General Report and Detailed Report, Vols. I-V. ETSU TID 4090 (1991).

River Wyre Preliminary Feasibility Study. Tidal Energy Barrage and Road Crossing Final Report. ETSU TID 4100 (1991).

Tidal Stream Energy Review. ETSU T 05 00155 REP (1993).

Wave Energy   A Review of Wave Energy. Thorpe, TW, (Volumes 1 & 2), ETSU R72 (1992).

Wave Energy: The Department of Energy's R&D Programme 1974-83, ETSU R26 (1985).

Technology Module for Shoreline Wave Energy Conversion in the UK. Martin, DJ. ETSU N 106 (1989).

An Assessment of the State of the Art, Technical Perspectives and Potential Market for Wave Energy. ETSU-OPET for DGXVII (1991).

The UK Wave Energy Resource. Nature 287. Winter, AJB. (30 October 1980).

Photovoltaic Systems   Review of Photovoltaic Power. Taylor, E. HMSO ISBN 0 11 413706 4. ETSU R 50 (1991).

A Study of the Feasibility of Photovoltaic Modules as a Commercial Building Cladding Component. ETSU S P2 00131 (1993).

Photoconversion   Photochemistry and Photoelectrochemistry Technical Assessment. Archer, MD. ETSU S-1327 (1992).

A Review of the Feasibility of Using a System Based upon Photobiology as a Large-Scale Source of Renewable Energy. ETSU S 1329 (1992).

Active Solar   Review of Active Solar Technologies. Halcrow Gilbert Associates Ltd with Stammers, JR. ETSU S 1337 (1992).

Passive Solar Design   Passive Solar Energy Efficient House Design – Principles, Objectives, Guidelines. Architectural Association School of Architecture. (In press).

Commercial Building Design. Burt Hill Kosar Rittelmann Associates/Min Kantrowitz Associates, Van Nostrand Reinhold (1987).

Energy Performance Assessment Reports. ETSU S 1160, S 1163.

Geothermal Hot Dry Rock   The 1990 Geothermal HDR Programme Review. Symons, GD. ETSU R 53 (1991).

Geothermal Hot Dry Rock, UK Government R&D Programme 1976-1991. ETSU R 59 (1992).

Conceptual Design Study for a Hot Dry Rock Geothermal Energy project – Main Report. ETSU G 153 (1991).

Geothermal Aquifers   Geothermal Energy - The Potential in the UK. (Ed. Downing, RA and Gray, DA). ISBN 0 11 884366 4 (1986).

Geothermal Aquifers: Department of Energy R&D Programme 1976-1986. ETSU R 39 (1986).

Geothermal Aspects of the Larne No 2 Borehole. BGS (1982).

Municipal and General Industrial Wastes   Energy from Waste: A Perspective. van Santen, A. ETSU. (1992).

The Combustion of Densified Refuse Derived Fuel. ETSU B 1205F. (1992).

Review of Municipal Solid Waste Incineration in the UK. WSL ISBN 0 85 624 4 (1990).

Industrial and Commercial Waste as Fuel, ETSU Technology Summary. TS 204. (1992).

Sewage Treatment Using Anaerobic Digestion, Mosey, FE. In 'Anaerobic Digestion' pp 205-235. Ed. Stofford, DA, Wheatly, BI and Hughes, DE, University College Cardiff. Applied Science Publishers Ltd (1980).

DoE Waste Management Paper Number 1 – A Review of Options. HMSO ISBN 0 11 752644 4 (1992).

Royal Commission on Environmental Pollution Seventeenth Report: Incineration of Waste. HMSO. ISBN 0 10 121812 5 (1993).

Landfill Gas   Power Generation from Landfill Gas. Ed. Gorman, JF, Maunder, DH and Richards, KM. ISBN 0 7058 1653 2 (1992).

National Assessment of Landfill Gas Production. ETSU Technology Summary 198 (1990).

Landfill Gas: Energy and Environment '90. (Ed. Richards, GE and Alston, YR). Proceedings of a Conference held in Bournemouth, England, 16-19 October 1990. ISBN 0 7058 1628 1 (1991).

A Basic Study of Landfill Microbiology and Biochemistry. ETSU B 1159 (1988).

The Guidelines for the Safe Control and Utilisation of Landfill Gas. ETSU B 1296 (1993).

Specialised Industrial Wastes

Scrap Tyres as an Energy Resource-a Review. Dagnall, SP. In Opportunities for Consumer Waste Recycling. Proceedings from IMechE Seminar 26 September 1991. ISBN 0 85298 782 X.

Poultry Litter as a Fuel in the UK – a Review. Dagnall SP. Proceedings from the 7th European Conference on Biomass for Energy and Environment, Agriculture and Industry. Florence, Italy 5-9 October 1992. P 05 07.

Hospital Waste Incineration – a Review of Progress in the UK. Proceedings from Solihull Conference, UK, 17 February 1993. ETSU Proc 2 (1993).

Large Scale Waste Combustion Projects: A Study of Financial Structures and Sensitivities. ETSU B 122 00291 REP (1993).

An Assessment of Mass Burn Incineration Costs. ETSU B R1 00341 REP (1993).

The Project Financing of Landfill Gas Schemes. ETSU B 7057 (1993).

Agricultural and Forestry Wastes

Anaerobic Digestion in the UK: a Review of Current Practices. ETSU B FW 00239 REP (1993).

Wood: A New Business Opportunity. An ETSU report for MAFF and the Forestry Commission. ISBN 0 7058 1689 9 (1993).

Wood: Energy and the Environment. ISBN 0 7058 1663 X. ETSU (1992).

High Capacity Whole Bale Straw Combustor. ETSU B 1257.

Sustainable Forestry: The UK Programme. HMSO 010 124 2921 (1994)

The Feasibility of Biomass Production for the Netherlands Energy Economy. NOVEM Project 71 140 0130 (1992).

Energy Crops

The Potential for Catch Crops and Native Vegetation as Fuel in the UK. Bradon, O and Price R. ETSU R 35 (1985).

The Potential of Miscanthus as a Fuel Crop. ETSU B 1354 (1992).

Short Rotation Coppice. ETSU B W5 00216 REP (1993).

The Public Perception of Short Rotation Coppice. ETSU B W5 00340 REP (1993).

Advanced Conversion Technologies

An Assessment of Thermochemical Conversion Systems for Processing Biomass and Refuse. ETSU B/TI/00207/REP (1993).

Sustainable Biomass Energy. Elliott and Booth. Selected Papers. Shell International Petroleum Company, Shell Centre, London SE1 7NA. (1990).

Printed in the United Kingdom by HMSO
Dd 297320, C30, 3/94, 59226, 567135